AUTOMOBILE ENGINES OF TODAY AND TOMORROW

Books by Irwin Stambler

BUILD THE UNKNOWN

A GUIDE TO MODEL CAR RACING

THE WORLD OF MICROELECTRONICS

THE WORLDS OF SOUND

SHORELINES OF AMERICA

AUTOMOBILE ENGINES OF TODAY AND TOMORROW

AUTOMOBILE ENGINES OF TODAY AND TOMORROW

BY IRWIN STAMBLER

A THISTLE BOOK
Published by

GROSSET & DUNLAP, INC.
A National General Company

New York

To Leigh Elisha Sprague

Copyright © 1972 by Irwin Stambler

Library of Congress Catalog Number 72-75786
ISBN: 0-448-21447-4 (trade)
ISBN: 0-448-26216-9 (library)

Acknowledgments

The assistance of many people and organizations in providing information and suggestions for improving this book's scope and accuracy has been invaluable. I would particularly like to thank Phil Geddes, formerly of <u>Commercial Car Journal</u> magazine, for checking the manuscript for technical accuracy. In addition, I am grateful for the help of such other individuals, companies or agencies as the Environmental Protection Agency; Ford Motor Company; Chrysler Motors; Battelle Memorial Institute; General Motors Corporation; American Gas Association; General Electric; McDonnell Douglas; Mazda Division of Toyo Kogyo; Westinghouse Electric; John Bold and the Garrett Corporation; Stanford Research Institute; Bill Lear and Lear Motors; Mack Trucks, Williams Research Corporation; Planning Research Corporation; Doug Diltz and the Carl Byoir Co.; National Aeronautics and Space Administration.

Irwin Stambler
Beverly Hills, California

Credits

The author and publishers wish to thank the
following sources for making their photos
and diagrams available to us:

General Motors: 4, 40, 51, 79, 80, 81, 94, 102, 105,
    106, 108, 109, 120, 126, 131, 136, 139, 141, 142
Ford Motor Co.: 5, 6, 7, 8, 9, 14, 15, 16, 17, 65,
    66, 69
Mazda Motors: 22, 24, 28, 29, 33
Garrett Corp.: 46
Detroit Diesel: 57
Mack Trucks: 60
Williams Research Corp.: 62, 70, 72
Lear Motors: 86, 87, 88, 89
McDonnell Douglas: 100
Battelle Memorial: 113
Irwin Stambler: 30

Contents

1

Updating the I.C.E.

The lifeblood of the modern world is power—power in the form of energy that runs the machines in factories, operates high speed computers, lights our homes and propels the amazing array of vehicles that take us over great distances in almost effortless fashion.

The availability of fantastic amounts of energy—electrical, mechanical, and chemical energy—at very low cost is so widespread in highly industrialized nations that it is taken for granted—unless something comes along to jolt our complacency. Man steadily plugged more and more electrical devices, from refrigerators to Hi Fi stereo sets and TVs, into those little oblong holes in the wall, until suddenly the power company couldn't meet the demand. A great city like New York, for example, had to tell its citizens to shut down air conditioners and unplug electrical units lest the entire system break down.

In the same way, we all have been guilty of taking the most crucial part of the automobile for granted. The prices of cars in the United States aren't low, by any means. But the fact remains that almost every person in the nation old enough to get a license usually can scrape together enough money to buy a used car that will work fairly well. The reason this became possible was the development of a compact source of power, called an automobile engine, that is truly a marvel of engineering know-how. Over several decades, the production engineers of this country came up with an engine design that could provide more power than teams of dozens or even hundreds of horses, at a cost per engine of only a few hundred dollars.

This engine was not achieved overnight. Many different experimenters and designers tried—and eventually discarded—all kinds of power plants before settling on a particular type called the internal combustion engine (I.C.E.). This engine offered many advantages: it could run on relatively cheap fuel, a "fossil" fuel* called gasoline; it had fewer operating problems than other engines; it could go farther without refueling or recharging than other engines; it could be built on a mass production basis for less cost than almost any other power plant.

For all of these reasons—and more—the internal combustion engine became the mainstay of one of the biggest industries in the world during the first two thirds of the 20th century. The automobile, powered by the mighty internal combustion engine, was the first giant step in "shrinking" the world. Each person or family now could have a metal chariot that could outlast any horse, travel hundreds of miles in a day, and free man from many tasks he himself used to perform painfully slowly.

---

* A fossil fuel is one that is made from the remains of once living plants or animals that have changed chemically over many millions of years to compounds that are easily burned to provide heat energy.

It isn't surprising that most people were happy about the automotive age. A good part of the country's population trooped to automobile showrooms each year to see the gleaming paint and chrome of the new models. The average person admired the increasingly trim lines of the latest cars, looked eagerly at changes in upholstery or the array of entertainment devices on the dashboard, but really took little notice of what was under the hood.

They didn't, that is, until scientists who study the makeup of the air on which man depends for life shocked everyone with one word—pollution. The air, they said, was becoming increasingly fouled with chemicals. These chemicals might, if unchecked, snuff out all life on this planet.

There were so many obvious sources of pollution that, at first, no one pointed a warning finger at the automobile. Most of the scientists who sounded the first alarms thought that cutting down the dangerous fumes that went into the air from huge oil refineries or large factory smokestacks would solve the problem. With public support, new laws were passed and industry had to spend millions of dollars to cut down on air pollution. Despite this, the air grew worse instead of better. In cities like Los Angeles and New York, it no longer took special instruments to detect air pollution. People could see ugly brownish-yellow layers hanging in the air. They could feel the effects of this "smog" on their bodies. On particularly bad days, people in smog-laden areas found their eyes watering and sometimes felt sick to their stomachs.

It was then that scientists realized there was a tremendous source of air pollution still uncontrolled—the automobile engine. True, each engine pumped only a very small amount of undesirable chemical compounds into the air. But multiplying these small amounts by millions of cars caused pollution totals in the tens of millions of tons per year. The automobile industry, like almost everybody—in-

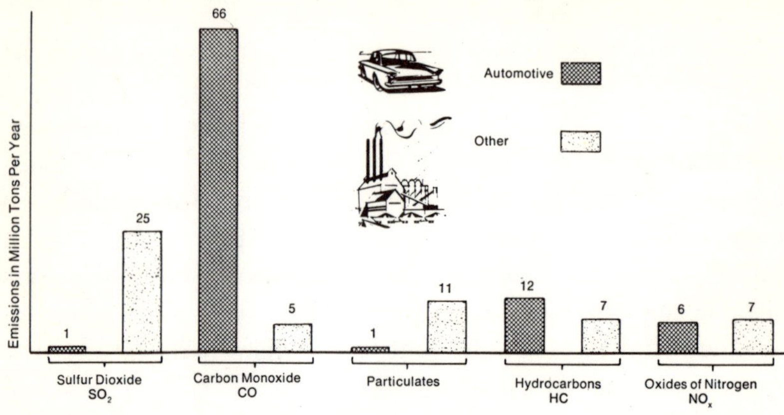

As this diagram of air pollution sources in the United States at the start of the 1970s indicates, automobile engines were prime sources of such dangerous products as carbon monoxide, unburned hydrocarbons and oxides of nitrogen. The survival of the internal combustion engine depends upon development of methods to drastically limit emission of such pollutants.

cluding the scientists—was caught off guard by the sudden discovery that the internal combustion engine, long considered a tremendous gain for mankind, now was a threat to our existence.

If this discovery hadn't been made, there would be little need for this book. The I.C.E. was such a fine machine for its purpose that no one in his right mind would have tried to eliminate it. But now that the dangers of pollution have been realized, the door has been opened for studies of different and new kinds of power plants. Principles once tried and discarded, such as the steam or electric motor, have been re-examined by teams of engineers and scientists both inside and outside the regular automobile industry. In addition, intensive studies have been started on completely new possibilities for automotive power, including gas turbines, fuel cells and hot-air-heat engines.

Automobile makers have developed advanced engines, such as the 351-4V CI (foreground) for the 1971 Ford Mustang, that incorporated many new pollution emission control systems.

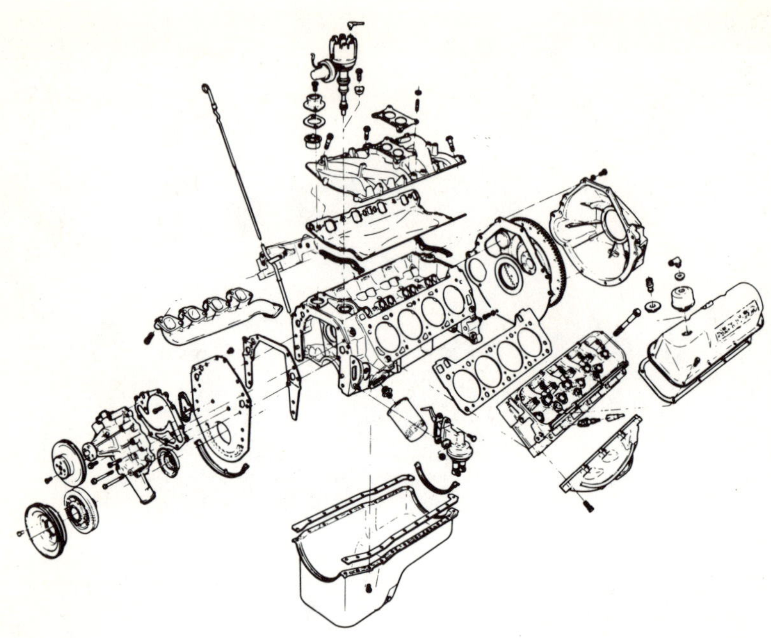

Exploded diagram of the 1971 351-4V CJ V-8 engine shows the main parts of a modern internal combustion engine.

Despite all this, however, the internal combustion engine has remained king. As all automotive experts know, this is still the best power plant for the average car on every basis except pollution. Development of other kinds of engines to the production level of today's internal combustion design can take many years. If laws were passed tomorrow banning the internal combustion engine in favor of some pollution-free type, the cost of each new engine would skyrocket. Lacking the years of patient development that went into the regular car engine, a competely new engine would be very expensive to produce. It would cost as much to buy a new car as to buy a twin-engine, executive airplane.

The obvious answer, then, is for Detroit engineers to find ways to drastically cut the pollutants emitted by internal combustion engines. To some extent, this already has

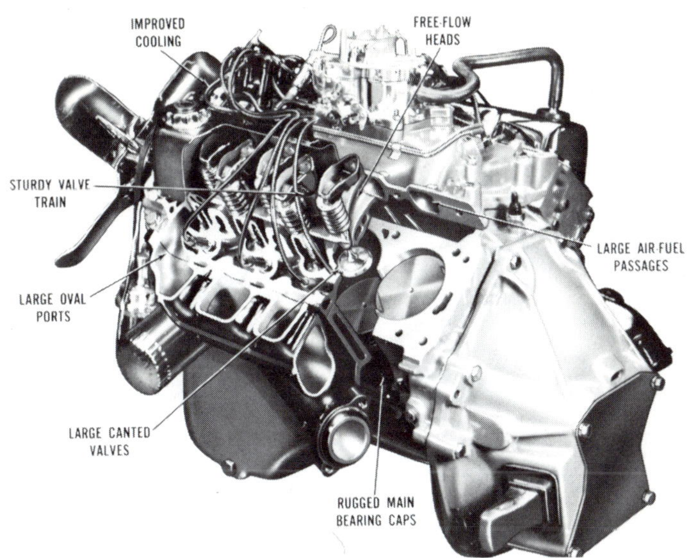

Cutaway model of Ford new 351-4V engine indicates some of the features of the 351-cubic-inch engine designed with the aid of computers. With a four-barrel carburetor design, the engine is rated at 300 hp at 5400 revolutions per minute. The two-barrel engine provides 250 hp at 4600 rpm.

been done by certain kinds of control mechanisms without adding too much to the engine cost. More important, other innovations now in research stages could cut pollution even more. But some of these require major changes in certain parts of the engine cycle.

To understand the problem and then look at possible solutions, the first need is a review of how the internal combustion engine works. It might be thought, looking at the seemingly confusing differences in tubing, metal parts and wires in the engine of one make of car and that of another, that the phrase "internal combustion engine" already covers dozens of different kinds of engines. The same impression might be gained from comparing 1972 engines with 1971 or 1970 designs, or comparing any of these with the seemingly simple automobiles of the very early days.

POWER STROKE   **EXHAUST STROKE**

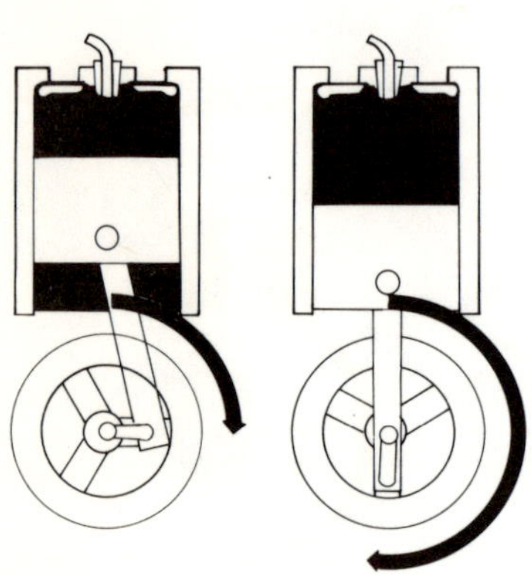

Figure 1

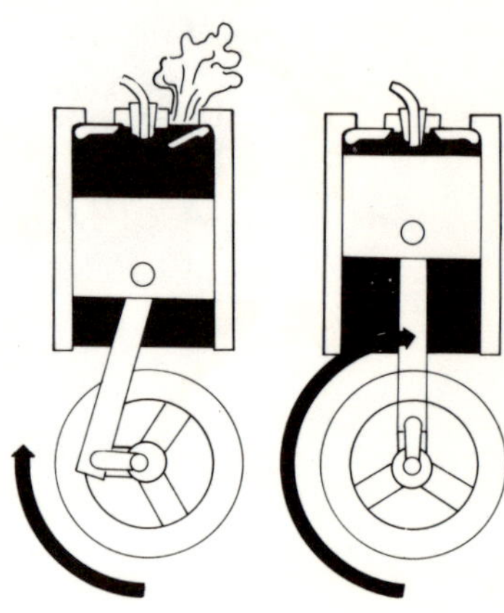

Figure 2

*The steps in a four-cycle engine operation (Otto cycle) are
shown in these diagrams.*

Actually, the basic principles of the modern internal
combustion engine are the same as those used in the first
successful gas engine. This was built by Nikolaus August
Otto and Eugen Langen of Germany in 1877. Their engine
was based on an engine described, but never built, by
French experimenter Beau de Rochas in 1862. The cycle,
or continuously repeated steps, by which all engines of this
long line work, has been named after only one of the many
people who contributed to the gas engine. This description
is the Otto cycle.

Internal combustion engines, typically, are either of the
two-stroke or four-stroke variety. That is, the four-stroke
system works with four up-and-down movements of a

INTAKE STROKE   COMPRESSION STROKE

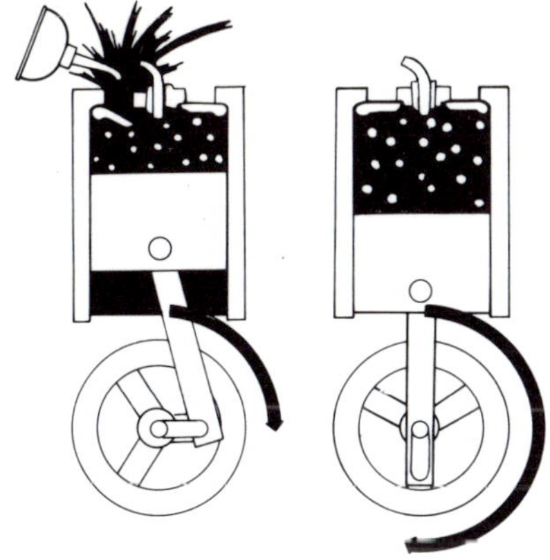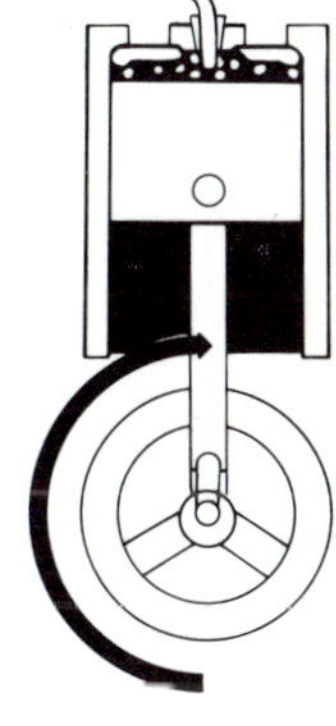

Figure 3   Figure 4

piston inside a housing called a cylinder. The two-stroke system is a method combining the four operations of the four-stroke cycle into two. It requires fewer parts but provides less useful power per pound of engine (at least for a car) than a four-stroke design. Because almost all automotive engines use the four-stroke cycle, this is the only one that will be discussed here.

The first step, after a gas engine has been started, involves preparing a mixture of fuel and air. This mixture is to be sent into a series of cylinders. In most cars, the mixing takes place in a carburetor, that is, a chamber which holds a measured amount of fuel from the gas tank, which is mixed with a certain amount of air. Some engines use the fuel injection method, in which fuel under high pressure is sprayed directly into the cylinders, combining with air sucked into the injector on the way.

At the start of the cycle, called the intake stroke, the

piston in a cylinder is drawn down to its lowest position, at which point one of two small valves in the top of the cylinder (or cylinder head) opens. A certain amount of fuel and air from the carburetor or injector is drawn into the cylinder. The piston then moves upward in the compression stroke, pushing the molecules of fuel and air into a very small space at the top of the cylinder.

Extending slightly into this part of the cylinder is a small metal tip from a device called a spark plug. The metal section is not continuous. There is a tiny separation between two parts of it called an air gap. These sections of metal are the ends of a piece of electrically conducting wire that extends back through the ceramic insulation of the spark plug to another part of the engine's electrical system, called a distributor. The distributor is a form of rotating switch. A contact in the distributor moves around, touching the metal wire ends from each of the cylinders. Every time such a contact occurs, a pulse of high voltage energy, supplied to the distributor from the car's high voltage coil, is directed through the metal wire into the metal tip of the spark plug.

The electrical current flowing into the plug then jumps across the small air gap to complete the electric circuit. When this happens, the energy in the current causes the fuel-air mixture in the top of the cylinder to catch fire. Burning under pressure, the mixture builds up a great amount of energy. To get rid of this energy, it expands, forcing the piston down on the third stroke of the cycle.

This is called the power stroke because it causes a metal connecting rod, attached to the piston, to transmit a force, through a series of bearings, to a crankshaft. The movement of the connecting rods in a series of cylinders is arranged so that the crankshaft rotates and provides the power (through the transmission system) to turn the car's wheels and make the car go.

It should be mentioned, briefly, that an important device called a flywheel is attached to one end of the crankshaft. Without the flywheel, there would be moments between one power stroke and the next, when the speed of the crankshaft might vary. The flywheel, though, stores up energy during the power stroke, then gives it up during the other strokes to make sure that the car will travel smoothly and at the speed regulated by the driver's foot pressure on the accelerator pedal.

After the power stroke is completed, the fourth step in the cycle occurs. This involves another upward movement of the piston in the exhaust stroke. This motion forces the burned exhaust gases upward and out through the second small valve in the cylinder head. (During the compression and expansion strokes, both intake and exhaust valves remain tightly closed.) The exhaust valve is lifted at just the right moment by a specially shaped, mechanical disk called a cam. This is attached to a rotating rod called a camshaft. The camshaft may be located in the crankcase or above the cylinders (the latter is part of what is known as an overhead cam engine), and it operates linkages such as push rods, tappets, etc., to open and close the cylinder valves at the proper times.

The exhaust gases in most internal combustion engines of the early 1970s simply flow into a manifold, a long chamber running the length of the engine. There they are collected from all the cylinders and passed into the exhaust line. The burned gases go through a muffler, a sound suppressor that cuts down on the noise of the rushing gas, and are ejected into the air through an exhaust pipe.

These steps are the ones that make the car run. But there are several additional systems needed to make sure the engine can *keep* running.

One of these is the cooling system. The cooling system is

vital because of the very high temperatures generated by the combustion of the fuel-and-air mixture. These temperatures can go to several thousand degrees Fahrenheit. In most cars, a liquid, such as water, flows through passages in the cylinder block and cylinder head, picking up heat from the metal walls surrounding the cylinders. The water, after taking on this heat, flows through a radiator. The radiator is a series of thousands of metal tubes that pass excess heat into the air while also cooling the liquid for its next pass through the cylinder system. A fan also is used to keep cool air circulating around the outside of the engine and to prevent the hot air from the radiator from collecting under the engine hood.

Equally important is the lubrication system. Oil is needed on many rotating parts, such as the push rods, the rotating linkages that move the cylinder pistons up and down, and the crankcase bearings. In a typical gas engine, the oil pan holding the oil is attached to the bottom of the crankcase. An oil pump forces oil over the surfaces that need lubrication. Otherwise, friction would cause moving parts to wear away or bind and stop moving. The function of the series of special rings around the top end of each piston is to prevent high pressure, high temperature gases from the combustion chamber from leaking into the oil section.

From the foregoing description, it's obvious there are many connecting parts that must be carefully fitted together to prevent leakage of fluids, gases, etc., from the engine. Improper assembly of some parts—such as the special gasket used to prevent fuel-oil mixtures from leaking out between the top and bottom halves of the engine block— can result in the leaking out of the high pressure fuel-air mixture. One way in which this happens is evaporation of fuel from either the carburetor or the gasoline tank.

The main source of pollution, though, isn't through

poor fittings, but from incomplete burning of the gasoline. This situation is a basic problem in an internal combustion engine. As chemistry teaches us, oxygen is required to produce combustion. The sizes of cylinders and the way in which the engine works make it impossible to force enough air into each cylinder with the fuel to insure complete burning. In other words, after all the available oxygen has been used up during the power stroke, there is still some unburned fuel left. This unburned fuel passes out into the exhaust system. In some cases, it manages to pass filters or gaskets and get into the crankcase. Unless something is done to chemically neutralize or somehow trap the pollutants inside the car system, they will go out into the air through the exhaust pipe.

The three most dangerous pollutants cited by scientists, before 1970, were hydrocarbons, carbon monoxide and oxides of nitrogen. In the early 1970s, another pollutant was added to the list—lead additives used to improve the efficiency of the fuel. As a first step in the battle to save the internal combustion engine, automobile engineers developed a number of new control devices to cut down on automotive pollutants.

One of the earliest devices was a control for "crankcase blowby." Studies showed that a great amount of unburned hydrocarbons got into the air from the road draft tube located under the car. This tube gets rid of excess air flowing through the engine system. Laboratory work showed that the air also picked up unburned fuel from the crankcase region. The answer, in part, was to add tubing to allow these blowby gases to flow back into the carburetor for re-injection into the cylinders.

A second series of devices was developed to cut down on the amount of unburned hydrocarbons released through the exhaust system. By 1971, two kinds of controls were

DUAL CATALYST CONVERTER SYSTEM

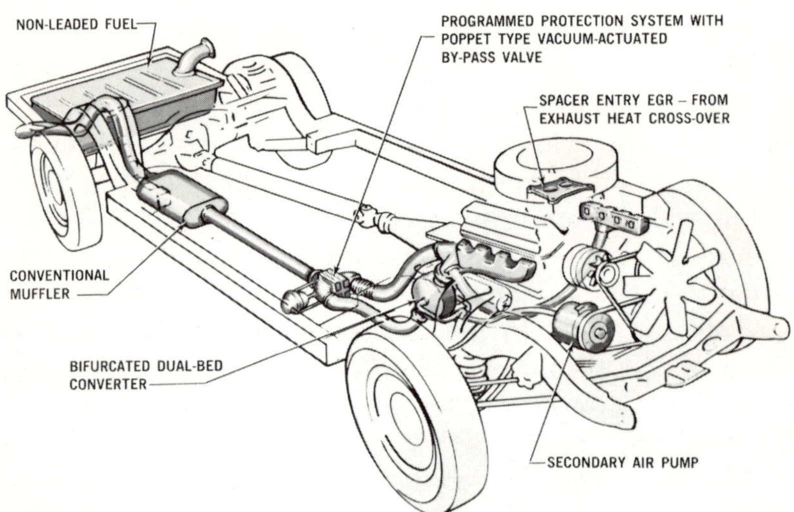

A dual catalyst converter system in a V-8 engine system. The chemical catalysts react with exhaust gases to change dangerous pollutants to harmless substances.

added to new car engines. One, called air injection, involves use of a high speed pump. The pump injects air into the exhaust manifold to "oxidize," or change the chemical makeup, of the unburned hydrocarbons before they enter the exhaust stream. The newly formed chemical compounds are not as dangerous to health as untreated hydrocarbons. The process, tests showed, also reduced the amount of carbon monoxide sent out through the car's tailpipe.

The second method is based on modification of the shape of the inside of the engine cylinders. This development, as General Motors Chairman J. M. Roche reported in 1969, was based on "an unusual series of experiments (in the research laboratories) using high speed photography in the engine's combustion chamber." These pictures

DUAL CATALYST CONVERTER SYSTEM
SIX-CYLINDER INSTALLATION

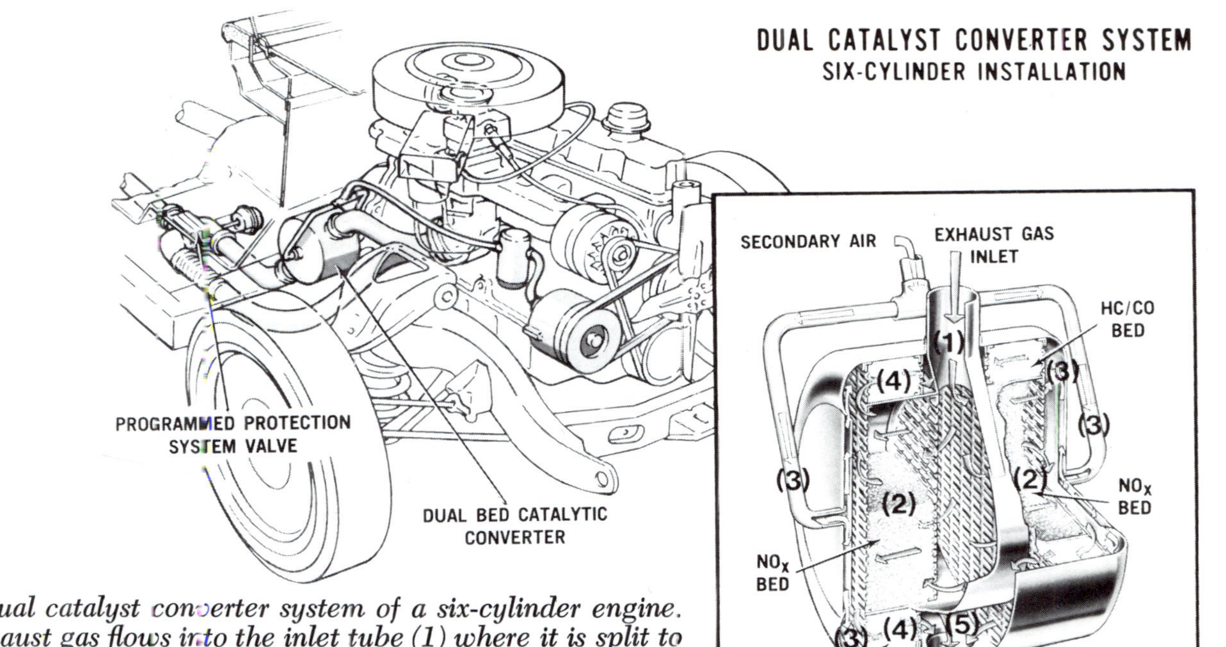

A dual catalyst converter system of a six-cylinder engine. Exhaust gas flows into the inlet tube (1) where it is split to run through two very large cylindrical beds (2) containing an oxides of nitrogen catalyst. It is then mixed with secondary air (3), then flows back through the second catalyst bed (4) to complete oxidation of hydrocarbons and CO. At (5), the gas is directed through an annular chamber to the outlet (6).

PROGRAMMED COMBUSTION
LOW EMISSIONS ENGINE

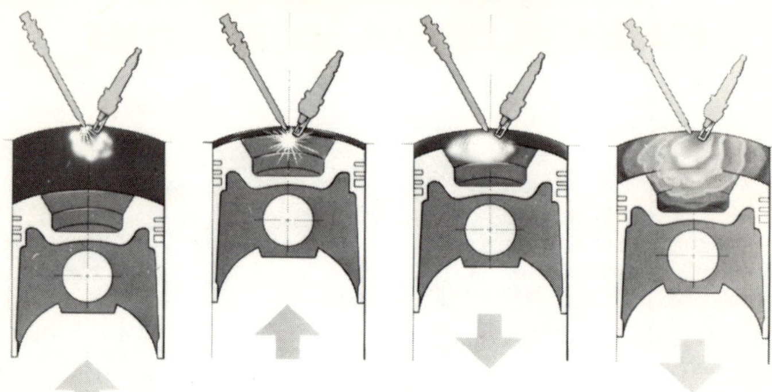

One new approach to lowering unwanted emissions from internal combustion engines is programmed combustion. In this system, mixing of fuel and air takes place inside the cylinders instead of in a carburetor. Air is introduced by a special injector in a swirling pattern that insures better mixing of fuel and air and more complete burning. The swirl permits the fuel-air ratios to be graduated from rich to lean outward from the point of injection. Note the specially cut out upper surface of the piston.

showed there was a "quench zone," that is, a zone of relatively cool metal on the cylinder chamber walls, "where unburned hydrocarbons try to hide." The approach taken here, called "controlled combustion," combined several changes to make sure most of these chemical compounds burned. One change was to shape the chamber in such a way that, after the fuel-air mixture is ignited, the hot gases expanding towards the chamber walls meet the minimum cooling or quench area. A second feature of controlled combustion is based on pre-heating the carburetor intake air. This extra heat energy produces more complete burning of the fuel in the combustion chamber.

The third technique is called evaporative controls. In this case, a small container of carbon, located near the carburetor, soaks up fuel vapor that would otherwise go

EVAPORATIVE EMISSION CONTROL SYSTEM

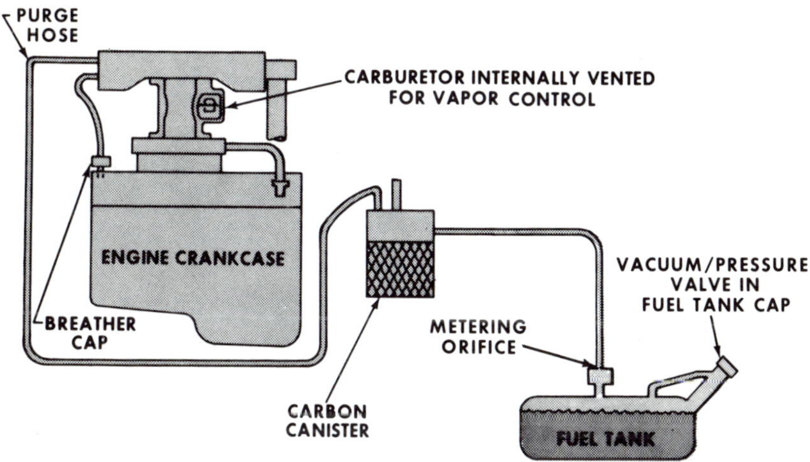

Evaporative control systems drastically reduced the escape of fuel system vapors into the atmosphere in cars of the 1970s.

into the air. At intervals, this vapor is discharged back into the carburetor for injection into the cylinders.

New engines with all of these controls on them eliminated a great amount of pollutants coming from car engines. The 1970 and 1971 engines, for instance, emitted only about 30% of the carbon monoxide and 20% of the unburned hydrocarbons of uncontrolled power plants. Assuming that most pre-1970 cars would be off the road in ten years, research scientists from motor car companies stated that the total hydrocarbon emissions in places like Los Angeles would be below the 1940 level.

Unfortunately, controlling hydrocarbons and carbon monoxide did not cut down oxides of nitrogen as well. In fact, the changes in the engine cycle caused by these control devices tended to increase oxides of nitrogen emitted by the new engines. In addition, scientists pointed out, the continuing rise in the total number of automobiles, trucks and buses in use in the years after 1980 would cause hydrocarbon and carbon monoxide levels to rise again, if there were no improvements over 1970 type controls.

Engineers and scientists knew that one way to cut pollutant levels of all compounds, from hydrocarbons to oxides of nitrogen, was to change the internal combustion engine from an "intermittent" type to a continuous combustion engine. There is a fraction of a second's time between the power stroke of one cylinder to the next when no burning takes place, due to the exhaustion of air in the cylinder. Tiny as these periods are, they contribute greatly to the total of unburned fuel collecting in the exhaust. Two new devices have been proposed which have the effect of making the internal combustion engine into a continuous combustion system. One is exhaust manifold "reactors," the other the addition of catalytic converters.

The "reactors" are really nothing more than very large exhaust manifolds. Such manifolds—two to four times the size of regular manifolds—have been tested in late model cars. The extra volume in the manifolds results in a large amount of exhaust gases remaining in the manifold which are heated to high temperatures. As these gases stay longer, the manifold temperature goes so high that by pumping in some extra air, the unburned fuel catches fire and continuously burns up.

Tests have shown that this method drastically lowers pollutant emissions. This process, however, generates very high temperatures of 2300 F. and more. Temperatures like these would turn ordinary manifold materials into liquid puddles in a few hours of operation. Thus expensive, high-alloy stainless steels and ceramic or other insulating coatings have been needed, which would make such a car engine very costly. In addition, because reactors work best with fuel-rich carburetor mixtures, they tend to raise fuel consumption by as much as 25%.

The second method is a little simpler. A catalytic converter is simply a housing containing special chemical catalysts that react with pollutants in the exhaust to turn the pollutants into harmless chemicals. This method has been studied in research laboratories for many years. It works, but it also has a number of vital drawbacks. One is that the best catalysts for the pollutants in question are precious metals, such as platinum. Some of the catalytic converters examined so far have been filled with metals whose total value is more than the price of the car.

There are some hopes that better, or at least much cheaper, catalysts can be found. However, most catalysts studied so far lose effectiveness rapidly because they become coated with other chemical compounds in gasoline. In particular, lead has been a problem, ruining most catalytic converters in a short period of time. However, lead has been of tremendous value to internal combustion engine operation up to now. It has been used in gasoline since 1922. At that time, scientists discovered that tetraethyl lead slows down the burning time of the fuel and eliminates the destructive "knock" that would otherwise occur in high compression engines. The addition of lead allowed the use of more efficient fuels and so cut the cost of operating the vehicle.

L. R. Hafstad, vice president, Research Laboratories, General Motors, points out that the cost of producing lead-free fuel of today's octane quality is 2¢ more per gallon. Using this figure to estimate the savings on the 1500 billion gallons of gasoline used by U.S. motorists from the early 1920s to 1969, a figure of $30 billion is reached. Today, savings from gas with lead added is estimated at about $1.5 billion a year.

On this basis, lead might never be removed. But scientific studies of the late 1960s showed that concentrations of lead in the atmosphere were beginning to reach a level high enough to bring the danger of lead poisoning to humans. So laws began to be passed in Congress to require, first, low-lead gas and, eventually, lead-free fuel. All of this made the catalytic converter look increasingly promising as an engine addition.

If all of these systems can't bring about a relatively low-cost, emission-free engine, tomorrow's cars may have to use some other power plants. This might be a radically different internal combustion design, such as the Wankel, or one of several other engines using an alternative to fossil fuel.

2

Rotary Revolution—the Wankel

At the start of the 1970s, the status of the internal combustion engine was becoming critical. The rising protest against the air pollution contributed by automobiles caused some lawmakers to introduce bills to outlaw the fossil fuel engine. These bills didn't pass, because it made no sense to outlaw the engines unless something was ready to replace them. Most of the legislators realized that the entire money structure of the U.S. would be turned into chaos if the use of the automobile ceased overnight.

Nevertheless, it was certain something would have to be done. These bills were a warning to the automobile industry. If it came down to a matter, literally, of life or death, the choice would have to be—no automobiles at all. The U.S. Government sought to insure a continuing automobile industry by providing funds for research, both in its

The Mazda R-100 sport coupe doesn't look much different from many foreign car imports of the 1970s. Under the hood, though, the car features the revolutionary rotary Wankel engine.

own agencies and in outside research groups, on smog-free engines. No one doubted that some kind of low pollution engine eventually would take over; the question remained, what principle would it be based upon?

In the internal combustion field, the spotlight suddenly turned to a radically different engine as the possible savior of the fossil fuel method. This engine, called the Wankel (pronounced Vahnkel) did not use the up-and-down movement of pistons to generate the forces needed to turn the car's wheels. Instead, it was a rotating combustion engine that had a roughly triangular-shaped piston with a hollow center. The piston rotated inside a housing with a pinched-in waist, looking something like the outline of a figure 8.

The reader might think the Wankel engine came to the fore because it was pollutant-free. Wrong! As of the late 1960s, the Wankel was considered far more "dirty" than the normal internal combustion engine. It might have been thought that the engine was highly developed, with few operating problems and ready for immediate widespread use by 1970. Wrong! As of the early 1970s, the car was still in its infancy and had a number of problems that still needed solutions to make it a true mass production design.

For all that, the engine had to have something in its favor. The large Japanese firm of Toyo Kogyo was producing and selling limited numbers of its Mazda cars powered by Wankels as early as 1967. Many foreign companies, Audi NSU Auto Union AG, Mercedes-Benz, Citroen and Rolls-Royce continued to study the Wankel principle, and some turned out limited numbers of cars for the commercial markets. And, to top it all, at the start of 1971, General Motors, the world's largest auto producer, signed a $50-million licensing agreement for the engine with Wankel

This array of rotary-engine-powered cars produced by Japan's Toyo Kogyo Company excited considerable interest among U. S. auto buffs and anti-pollution fighters in 1971.

patent holders, Wankel GmbH and Audi NSU, and the U.S. licensee, Curtiss Wright. The agreement was that GM would pay the other firms $50 million over a five-year period as long as it continued development work on the principle. If GM did complete this payment, it could then manufacture Wankel designs without having to pay any further licensing fees.

Obviously, the Wankel principle is a very important one. Examination of the system showed engineers that its pollution problems seemed solvable. In fact, by 1970, work by Japanese and German firms already had lowered the unburned hydrocarbon levels considerably. More important, it became apparent that such emissions as carbon monoxide and unburned hydrocarbons could be cut *without* increasing the oxides of nitrogen, as was the case with conventional internal combustion engines. A primary attraction of the Wankel, it turned out, was that it operated, even in early development stages, with a minimum emission of oxides of nitrogen. By the end of the 1960s, it had become apparent that oxides of nitrogen was a greater villain in smog than hydrocarbons, a fact previously overlooked.

Actually, from a performance standpoint, the Wankel principle had long made more sense than regular car engines. It is basically much simpler, requiring far fewer parts, and can operate much more quietly. When some of the bugs in the early designs are worked out, it also promises to run trouble-free for considerably longer periods of time than the pre-1971 engines. It also costs less to repair if something does go wrong.

However, with all these advantages, the Wankel did not make much of a dent in the automobile field during the mid-1960s. It was simply that at that time auto builders knew they had a proven, low cost engine that did the job.

They could see no reason for spending the many tens of millions of dollars needed to perfect a new kind of engine —until the demon of air pollution forced their hand.

One man had confidence in the great potential of rotary combustion—Felix Wankel, although he gave as little thought to air pollution as did anyone else way back in the early post-World War II years. Wankel had been an expert on rotary combustion machines for many years and felt sure they had much to offer. At one time or another, other inventors and engineers had thought so too, but the problems of designing a practical rotary machine had defeated all of them. The main attraction of a rotary engine is that it simplifies the process of turning the chemical energy of burning fuel into the rotary energy that eventually spins the wheels. In the engine described in Chapter I, the up-and-down motions of a series of pistons have to be changed into rotary motion through linkages to the crankshaft. In the rotary type engine, the piston itself provides the rotating motion directly to the transmission.

Wankel had been in charge of a laboratory that had been destroyed as a result of World War II. In 1951 he established a small new workshop inside his own home in West Germany, with the prime goal of developing a successful rotary combustion engine. As Wankel later stated, he had found there were 864 different methods which could be divided into four major classifications. Of these, his analysis had shown 278 to be unworkable, 149 to be workable and the rest remaining to be analyzed or invented.

After spending hour after hour, day after day, considering all of the possibilities, Wankel came up with a theory that looked as though it could be the basis for a practical system. He reached this point in 1954, after almost three years of dogged research partly supported by NSU Motorenwerke AG, a company then concerned with building motorcycles rather than cars. Wankel's discovery was that he could reproduce the four-stroke cycle of his earlier countryman, Dr. Otto, by rotating an equilaterally shaped triangular piston in a specially shaped housing that also rotated. The completed engine was much too complicated for immediate application. But it dramatically demonstrated the feasibility and promise of Wankel's ideas.

The second breakthrough that made the Wankel engine a contender for vehicle use came from NSU. Dr. Walter Froede, who had followed Wankel's work closely, realized a more practical approach was needed. He decided to concentrate all rotation within the envelope of the piston, allowing the housing to remain stationary. He did this by making the center of the piston a hollow internal gear whose teeth meshed with a smaller, eccentric gear. Froede's system added more vibration than Wankel's almost vibration-free design, but it eliminated a great many extra parts required in the earlier engine.

The combined Wankel-Froede system has served as the basis for all of this family of engines built since that time, though there have been variations in some detailed parts.

Using the Mazda engine as an example, the general operation of the Wankel engine involves rotation of a triangular rotor with arched sides inside the epitrochoid*-shaped chamber, which has flat walls. The combination of gearing is designed to keep the three apexes of the triangular piston constantly in contact with the chamber sides. The gear system consists of three main parts: a rotor gear; a fixed gear, attached to one side of the housing, which engages the rotor gear; and an eccentric shaft that passes through the center of the fixed gear. The shaft, which is the output shaft, is supported by main bearings in the front and rear sides of the housing. The gear ratio is de-

---

* The mathematical name for the inside shape of the housing.

Opened Mazda engine cylinder shows the three-apex piston that revolves within the epitrochoid walls of the housing.

signed to provide a one-to-three relationship between the rotor and the eccentric shaft. In other words, for each complete rotation of the rotor, three rotations of the shaft take place. The rotor is in sliding contact with the eccentric shaft. It is this contact that makes the shaft rotate to provide the forces needed to turn the car's wheels.

The rotating movement of the piston is such that each of the three apexes of the triangle follows an oval-shaped curve. When this happens, as Mazda's engineering manual points out, "three operating chambers, which show smooth changes in volume on a uniform cycle, are formed between the arched sides of the rotor and the inner epitrochoid wall of the rotor housing."

The volumes of these three chambers change greatly, depending on which part of the four-stroke Otto cycle is occurring. The intake stroke takes place when the chamber volume facing the intake port has its minimum space

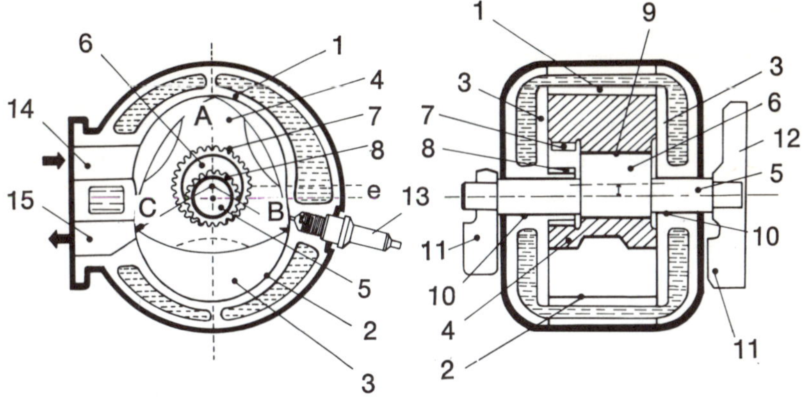

1: Rotor housing   2: Epitrochoid   3: Side housing
4: Rotor   5: Eccentric shaft   6: Eccentric part   7: Rotor gear
8: Fixed gear   9: Rotor bearing   10: Main bearing
11: Balance weight   12: Flywheel   13: Spark plug
14: Intake port   15: Exhaust port
A.B.C.: Apexes of the rotor   e: Eccentricity

Main parts of the Mazda (Wankel) cylinder. Sketch at left is a view looking from the front of the housing; the right-hand sketch is a side view of the same housing.

volume. At this point, the air-fuel mixture from the carburetor is drawn into the chamber. The flow of fuel and air continues as the piston rotates, gradually increasing the volume available for the mixture. The steps in the intake operation, as indicated in the sketches, are numbered 1-4 with the fourth step being the one in which the maximum volume is available for the fuel and air.

It should be noted that the four steps of the Otto cycle are not separated as is the case with a reciprocating engine. Because of the piston's rotation, the three other steps are going through various phases even as the intake step is occurring in one of the three chamber sections. Thus, while the intake process is going on in phases 2 and 3 of the sketch, the compression phase for the previously inserted fuel-air mixture is going on in the next area formed by the

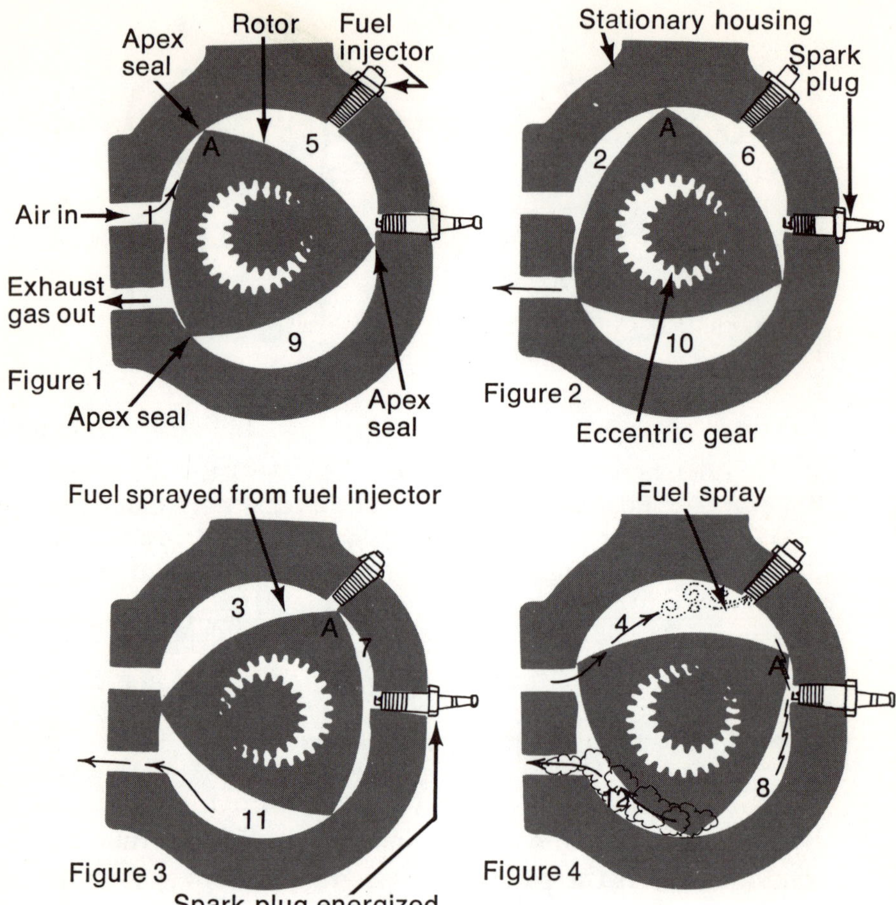

In the operation of the Wankel engine the action starts with the intake of air (1). The intake process continues through steps (2), (3) and (4). Fuel is mixed with the air in steps (3) and (4) as the rotor apex marked (A) continues its clockwise motion. Going back to step (1) the compression sequence begins in (5) and continues until fuel and air are squeezed into the minimum volume in (7). At this point the spark plug ignites the mixture. The force generated by the explosion acts to rotate the piston in the power stroke phase, steps (8), (9) and (10). As the rotor completes its revolution, the exhaust products are swept out of the exhaust port in steps (11) and (12).

second arched side of the piston and the chamber wall. As the diagrams show, while the intake stroke starts, the piston is beginning the compression stroke on the fuel and air from the previous intake stroke. This compression stroke can be followed through phases 5, 6 and 7 in the top left and the two right-hand diagrams. By phase 7, maximum compression has occurred and the mixture is ready for ignition. At this point, the spark plug fires, causing the mixture to start burning.

The power, or expansion, stroke follows. The burning gases push against the piston, steadily increasing the volume on the left-hand side. During this step, of course, rotating motion is imparted to the shaft. During all these phases, the contact of the piston apex separates what is going on in one area from the actions in the other two areas. As the expansion stroke continues, the apex at the top of this piston side moves around until it begins to uncover the exhaust port. The next apex then acts like a broom to sweep the burned-out gases from the chamber and out through the port in the exhaust stroke.

The preceding description underlines many of the advantages of the Wankel. The discussion of the way fuel and air are injected and the combustion products later exhausted indicate how this is done without requiring valves. Eliminating the valves also does away with the need for all the complicated mechanical gadgets used to make the valves open and close. The Wankel doesn't need cams or camshaft, or flywheels to smooth out the jerky motion of the piston, nor does it need all the linkages from the pistons to rotate a massive, hard-to-form crankshaft.

As the description of the Wankel cycle shows, there is one third of a revolution of the rotor for one revolution of the eccentric shaft. In addition, there is one ignition spark generated for every Wankel rotor revolution. By comparison, the Wankel cycle, therefore, has twice the power impulses of a reciprocating (moving to and fro) piston four-stroke cycle. Put another way, the displacement volume of the Wankel does two times the work of the volume of the conventional 4-cycle system.

Combining this factor with the relative simplicity of the Wankel, this new kind of internal combustion engine turns out the same amount of power as an equivalent reciprocating engine, but has only half the weight and half the volume of the conventional power plant. There are indications that continued improvement of the Wankel concept could reduce its size and volume, compared to comparable reciprocating engines, by an even greater percentage.

The compactness of the Wankel therefore opens new paths for automobile designers to follow. Much more space can be devoted to storage and passenger comfort. Assuming that all the problems of a low cost, mass production model can be solved, automobile experts predict the Wankel will cause a virtual revolution in body styling, riding comfort and overall car performance.

The one "if," as has been stressed, is overcoming a number of drawbacks. The most important, at present, is exhaust pollutant emissions. There are other things as well. One vital problem for long life and good performance is development of special seals, particularly for such hard-to-seal places as the three piston apexes. Spark plugs also tend to have shorter lives in the Wankel for several reasons. One reason is their location at a point where the fuel-air mixture is more highly compressed than in a reciprocating design. This causes increased heat that offsets the cooling from the air in the incoming mixture. Secondly, spark plugs in the Wankel fire more often in a given period of

Interior view of the dashboard and stickshift system used in the Mazda RX-2 rotary-powered model.

time compared to conventional I.C.E. designs. In addition, as of the early 1970s, the Wankel tended to use as much as 10-15% more fuel than a comparable reciprocating engine.

Looking first at the emission problem, the Wankel still has the problem of any internal combustion engine. It's not possible to inject enough air to insure complete burning of the charge of gasoline. Besides this, it also has the problem of chamber wall quench. That is, the relatively cool walls provide a hiding place for some hydrocarbons from the burning process. The shape of the piston and the housing also result in a relatively higher surface-volume ratio compared to the reciprocating engine. As was pointed out in Chapter I, the best way to minimize unburned hydrocarbons in the combustion chamber is to reduce this ratio.

This combination of a decrease in the total amount of fuel-air mixture burned in the Wankel chamber, plus sealing difficulties, resulted in an output of twice the unburned hydrocarbons than reciprocating engines in early versions of the Wankel. By the early 1970s, though, a number of steps had been taken to improve things. For example, engineers in Europe and Japan examined ways of widening the piston and housing to increase the ratio. In addition, they increased the chamber volume by designing the arched sides of the piston with depressions in them. Obviously, though, the amount of improvement in surface-volume ratio gained in both these ways was limited. Depressions could only be made so deep because the piston was hollow. On the other hand, making the piston and housing much larger would defeat the advantages of light weight and small volume of this kind of engine.

The major approach to minimizing unburned hydrocarbons was one already being studied for reciprocating engines—use of an exhaust reactor. A variety of reactors seemed feasible by the early 1970s, including one under test in a joint University of Michigan–Curtiss Wright program directed by engineering professor Dr. David Cole. The reactor, as in the reciprocating type, is a large volume chamber where high temperature air is injected to burn any leftover fuel. Cole's tests showed, as did others in other research groups, that the Wankel system had a number of advantages for the reactor approach compared to conventional piston types. The exhaust gases from the Wankel are much hotter, which is very desirable for the afterburning process. In addition, the Wankel reactor can work well with lean fuel-air mixtures, whereas the reciprocating engine of the '70s had to use extra fuel to make its reactor practical.

Improvement in emissions achieved by Cole's group was

gained not only from the reactor method, but by variations in other things, such as combustion chamber shapes and intake and exhaust port shapes.

The goal of work such as Cole's is to drastically reduce hydrocarbon emissions below the minimums set by the anti-pollution laws of the early 1970s. Meanwhile, reactors built for production engines of the period, such as the Mazda or the NSU Ro80 engine, already have reduced the levels significantly.

These engines start with an advantage. Studies have shown that while carbon monoxide and unburned hydrocarbons result from incomplete burning, the nitrogen pollutants result from chemical reactions in the engines. Very high temperatures cause the reactions that form these nitrogen compounds. The higher the *peak* temperature (not the average temperature), the greater will be the amount of nitrogen oxide pollutants produced. The reciprocating engine ends up with exhaust gas temperatures below that of the Wankel chamber. But, during the cycle, the peak temperature reached is slightly above the peak in the Wankel. Thus the amount of oxides of nitrogen coming into the Wankel exhaust manifold is lower than that in the reciprocating design.

Though all engine areas needing sealing require careful study, the most critical problem for the Wankel has been the apex seals. These seals have to maintain almost perfect contact with the housing so that gas in the chambers where compression and expansion steps take place doesn't leak into other sections. If this happens, the thermal efficiency and the power output of the engine can drop sharply. The relationship of the apex to the housing sides changes during rotation, due to the irregular shape of the housing. Therefore the apex seals must be designed from a material that can withstand very high temperatures, and also be capable of a certain amount of in-and-out motion to insure continuous wall contact.

The temperatures of well over 1,000 degrees rule out such materials as rubber or other flexible materials. Instead, the seal must be made of such things as ceramics, carbon compounds or piston-ring iron with some form of spring forces to keep them pressed against the housing wall. The difficulties in developing such a seal were discovered early in the NSU program, when special carbon apex seals failed and had to be replaced after only 10-15,000 miles of travel. By the early 1970s, however, seals that had reasonably long lifetimes had been developed.

It wasn't easy to do. A seal had to be developed that would not break under operating conditions. Also, the design had to be one that would not dig into the housing. For a long time, promising seal designs had to be rejected because they caused chatter-marks on housing walls.

Says Mazda, "Chatter-marks are wave-like marks occurring on contacting metals caused by vibrations. On the rotary engines, these marks appear as wave-like wear on the trochoid surface of the housing when the seals on the tips of the rotor slide against this surface." This condition not only impaired engine performance, but also shortened engine life.

"This was an extremely difficult problem which could not be overcome by merely working on ordinary seal materials or treating the trochoid surface in the ordinary way. It was necessary to test and try every possible material available on this earth. Day in and day out, tests on the chatter-marks were continued on the benches. Heaps of rotor housings, all with these marks, built up in our laboratory."

Mazda had to develop special new instruments to measure the vibrations, and process the data on computers to find new seal arrangements to overcome them. Finally, after six months, the research bore results. It was found that vibration levels could be drastically reduced by drilling one hole crosswise near the tip of the seal and then drilling a second one lengthwise to intersect with the first one. These holes redistributed the forces on the seals so that vibration was almost eliminated. The seal material developed for the production engine was a special blend of carbon and aluminum.

This work, together with many other advances in rotary combustion design, helped make the Wankel a usable engine by the early 1970s. Not all the problems had been solved, though, and the engines used in the cars of this period were basically hand-crafted rather than mass-produced. The Wankel was definitely on the scene to stay, but there was still plenty of competition from other kinds of engines for the future markets, particularly for heavy duty uses such as trucks and buses.

3

Brute Power—the Diesel

One of the favorite songs of truck drivers in the late 1960s was "Diesel by the Tail." The song sold over a million dollars' worth of records for the country-and-western team of Jim and Jesse, and was often heard coming from the juke boxes in the roadside eating places favored by the men who ride the cabs of massive cargo-carrying vehicles.

It is no surprise that a song with this kind of title should find an audience with trucking people. The brute force of the diesel engine has long been the mainstay of large transportation systems. As of the early 1970s, the diesel was the preferred engine not only for trucks of all kinds, but also for buses, large tractors and earth-moving machinery, and for modern railroad locomotives. The diesel principle also found use in some of the largest power plants built for stationary systems, including huge engines for electric power-generating plants or for operating rock-crushing machines in mining operations.

The diesel is a member of the I.C.E. family. The basic steps in its cycle closely resemble those described for the gasoline engine of Chapter I. However, it uses a different method of igniting its fuel-air mixture to drive its piston. This method provides the diesel cycle with much better heat efficiency (percentage of heat energy turned into useful power) than the spark ignition method. In fact, the efficiency of the diesel at this writing is one of the highest of any type of engine. The average heat efficiency is 34-36%. By comparison, the thermal (heat) efficiency of the best gasoline internal combustion engine (piston type) is about 25%, that of typical steam and gas turbine engines about 20%, and reciprocating steam engines about 12%.

The diesel has a high efficiency, which means it requires less fuel to go a given distance compared to the others. It also can work with a number of different kinds of fuel, including relatively cheap low grade gasoline or kerosene. These economic advantages made the diesel the favorite for many industrial-commercial uses.

Of course, the question arises, why isn't the diesel widely used in passenger cars if it's so efficient? There are several reasons. One is that the diesel must be carefully serviced to maintain its performance advantages. The engine parts must be constantly checked for proper alignment. A small variation in the desired clearances for valves and pistons, for instance, can result in a large loss in thermal efficiency. Even more important, without regular overhaul by first rate engine mechanics, the engine can emit far more of some types of pollutants into the air than a very "dirty" conventional passenger car engine.

Obviously, only large organizations or industries can

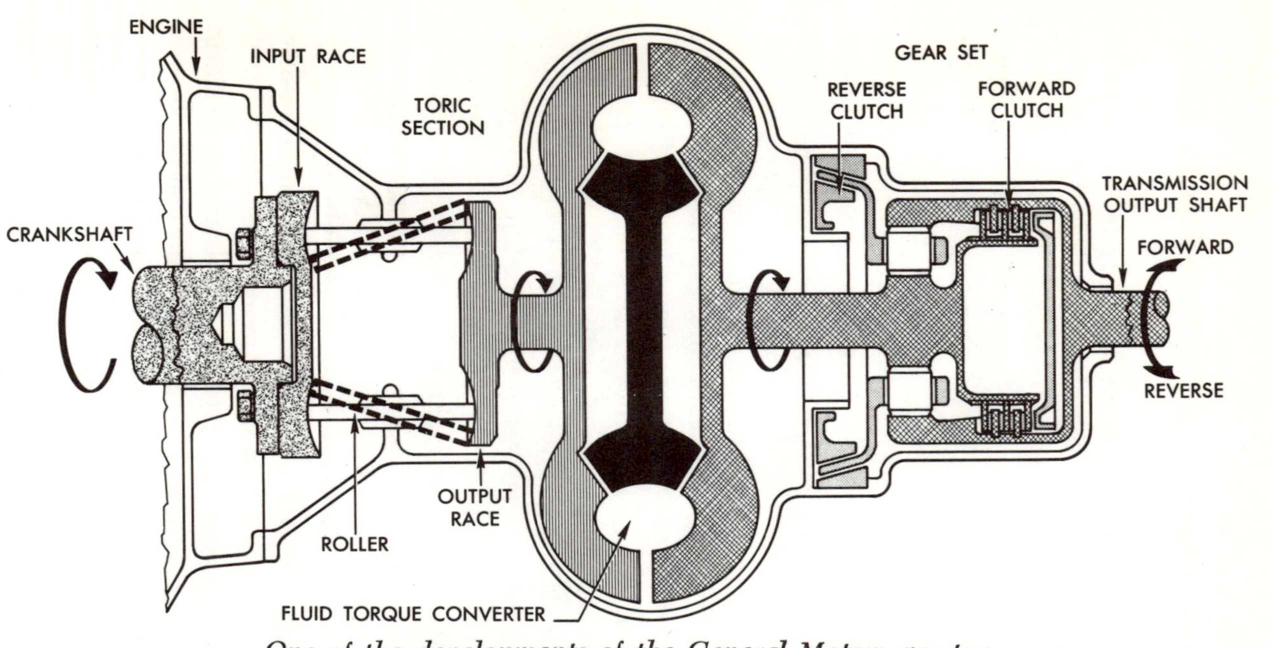

One of the developments of the General Motors gas turbine engine program was an advanced type of transmission based on the use of adjustable toroid-shaped surfaces to transmit power from the engine crankshaft to the car's wheels. This toric transmission, the main parts of which are shown here, was later used in the SE-101 steam car.

support the kind of instruments and highly skilled mechanics to provide the required diesel maintenance. Either that, or the passenger-type car must be sold at a relatively high price to provide a profit margin that will make it worthwhile for the car maker to maintain extremely capable service shops. Thus diesel-powered Daimler-Benz, Mercedes-Benz or Peugeot passenger cars are on the market.

Another drawback to passenger car use of the diesel is its high "specific" weight. That is, for each horsepower produced, a typical diesel engine must be bigger and heavier than a gasoline engine. The best ratio of engine weight to horsepower for a four-stroke cycle diesel is about ten pounds, roughly double that of an average gasoline internal combustion design. Diesel car engines, which use the somewhat less efficient two-stroke cycle, get the ratio down to six pounds per horsepower, but good gasoline engines can be made with lower ratios.

The additional engine weight of a diesel is no major problem where heavy duty vehicles are concerned. The engine weight is only a small percentage of the total vehicle weight, and the gain in poundage compared to a gasoline engine is more than offset by the lower cost of operation of the diesel.

The history of internal combustion engine development has been closely associated with Germany, and this pattern holds true for the diesel as well. The man who conceived the idea for this power plant was Rudolf Diesel, a German engineer. He was born in Paris in 1858, but his parents were German and he spent most of his life in Germany.

He started in the refrigerating field, but soon settled in Munich to concentrate on developments in heat engines. He felt there might be a better way to develop power than the spark ignition methods of Otto. Eventually Diesel came up with the idea of generating the high temperatures needed to explode a fuel-air mixture from compression alone. In 1893, he outlined his theories in a paper titled "The Theory and Construction of a Rational Heat Motor," which included a description of a diesel oil machine he had patented in 1892.

Diesel gained support from the Krupp and Augsburg machine factories, who turned his plans into a working engine by 1897. Visitors to the 1898 Munich fair were impressed with the first engine of this kind ever shown publicly. In particular, Adolphus Busch, of the St. Louis family that founded the Busch Brewing Company, was excited enough about the engine to pay Diesel a million gold marks for U.S. production rights. Back in the United States, Busch founded the Diesel Motor Company of America—later called the American Diesel Engine Company—which manufactured the first diesel engine in the world to be successfully used commercially. This engine, however, was installed in an electric power-generating system rather than in a moving vehicle.

Diesel himself outlined the working steps in his engine at a scientific meeting in Paris in 1900, and the basics have not changed very much up to the present:

"1. A suction stroke during which air alone at atmospheric pressure is drawn into the cylinder.

"2. A compression stroke in which this air is compressed to a pressure of 500-600 lbs. per square inch. (This is the heart of the system. The goal is to compress the molecules of air into so small a space that tremendous internal energy builds up. Literally, the air is pushed together until it is red hot.)

"3. From the end of the compression stroke, and for a

short period after, regulated admission of the fuel takes place in the form of a fine spray in such manner as to cause combustion to occur at (approximately) constant pressure.

"4. Expansion of the ignited mixture of fuel spray and air takes place during the remainder of the working stroke.

"5. Expulsion of the burned products during the next stroke."

It can be seen that, except for the way the fuel-air mixture is ignited, the description covers a four-stroke cycle almost identical to the four-stroke Otto cycle. As in the gasoline engine, the diesel uses a number of cylinders so that the motions of the pistons are transmitted through linkages to rotate a crankshaft. The diesel is simpler than the conventional gasoline engine because it doesn't need an external ignition system.

The diesel can be designed in a two-stroke cycle as well as a four-stroke one, as is the case with a gasoline engine. In both cases, the two extra strokes are eliminated by combining operations that take place separately in the four-stroke method. This is done through the use of an approach called scavenging. In scavenging, a low pressure source of additional air is used to blow fresh air into the cylinder which simultaneously scavenges—forces out of the chamber—the used up, burned gases.

In a two-stroke cycle, as the piston moves downward during the intake stroke, it uncovers two sets of ports at the same time, one for exhaust and one for air intake. As soon as the ports are uncovered, fresh air blown in from one side forces used gases out the other. The rotating linkage attached to the bottom of the piston then moves it back up past the ports, simultaneously cutting off air inflow and exhaust while compressing the air into a small volume at the top of the cylinder. When maximum compression is reached, an injector sprays fuel into the air, whose temperature by now is 1,000 degrees or higher, causing spontaneous ignition. The gases expand, forcing the piston down in the power stroke until the ports are exposed again.

Theoretically, accomplishing a given rotation of the crankshaft in two moves of the piston rather than four would provide for a more effective power plant. If a perfect two- and four-stroke cycle of the same size and speed could be built, the two-cycle engine would develop twice the power of the four-stroke system. It turns out, however, that the two-stroke loses efficiency from several areas. The most important problem in two-stroke is the need to rapidly clear out the burned gases from the cylinder. If simple, open ports are used, the scavenging action is incomplete, leaving waste products in the chamber when the next cycle is in its compression phase. In this case, the efficiency of the two-stroke cycle goes down considerably.

An improvement in the two-stroke engine-scavenging operation can be gained by using exhaust and intake valves rather than simple ports. The 280-horsepower truck diesel made by Cummins Engine Company, one of the most widely used engines, was designed, therefore, with valves even though it is a two-stroke system. However, going to valves cuts into one of the main theoretical advantages of the two-cycle engine—its simplicity—by adding more mechanical operating parts.

Even with valves, a well-designed two-cycle engine usually can provide the same power output for less weight than a four-stroke design. This relationship becomes increasingly important as the required engine size increases. Actually, for smaller four-cycle engines, variations in the design approach can make up for the theoretical advantages of two-stroke. For passenger cars, in fact, the total efficiency (made up of a combination of power output and heat efficiency) of a good four-stroke diesel is better than a good two-stroke even though the two-stroke has the better thermal efficiency in this size range.

The result is that almost all very large diesel systems are two-stroke types, and small ones are four-stroke. The medium size range covers some truck sizes; and models can be found using either the two- or the four-stroke cycle, depending on the particular performance requirements for the vehicle.

From a pollution standpoint, a finely tuned diesel offers some impressive advantages over a gasoline internal combustion engine. The reason is the excess of air used in the diesel cycle. The diesel does not use a carburetor as does a gasoline engine, and it begins with a cylinder filled only with air. What this means is summed up by the *Encyclopaedia Britannica* in this way: "The proportion of air to fuel, called the air-fuel ratio, is constant for gasoline engines in the ratio of 14.5:1; that is, 14.5 pounds of air to one pound of fuel." Because of the way the diesel accomplishes its air and fuel mixing, the article points out, "The diesel always compresses the same amount of air while the amount of injected fuel is varied to suit the load conditions. Thus the air-fuel ratio varies from about 22:1 at full load to about 85:1 when the engine is idling. The diesel engine requires an excess amount of air as compared to the gasoline engine."

For this reason, the exhaust from a diesel without any special anti-pollution controls, contains much less carbon monoxide and a sizable reduction in unburned hydrocarbons. Unfortunately, the amount of oxides of nitrogen isn't much different. The reason, as noted in earlier chapters, is that formation of these oxides comes from high

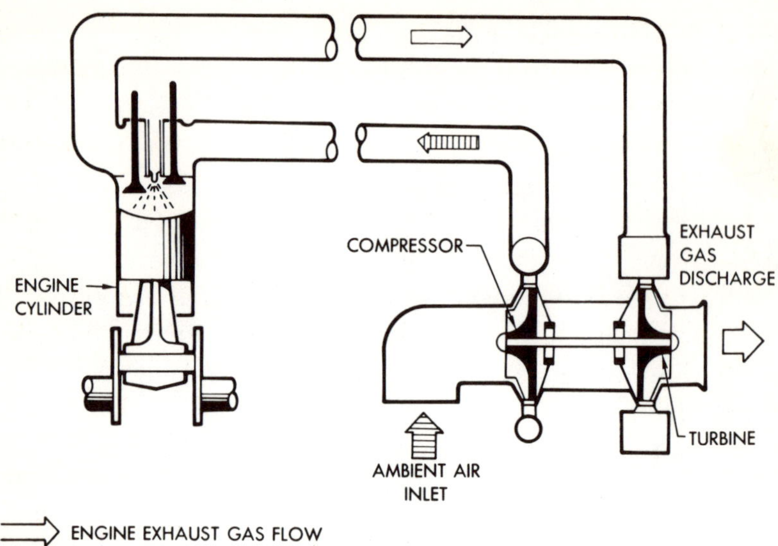

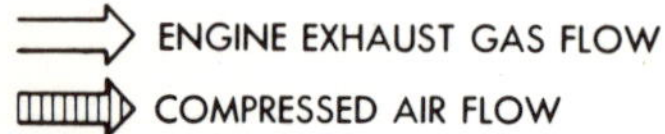

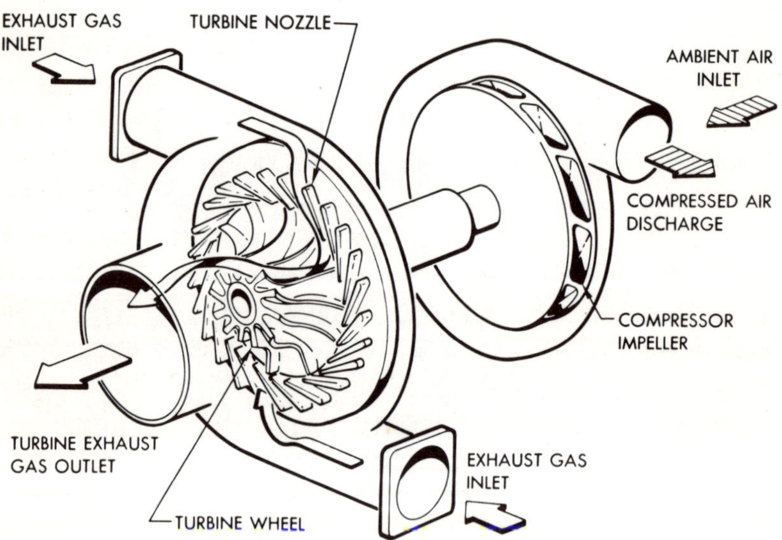

Main parts of the Garrett Turbocharger, used to improve operating performance of a diesel engine.

temperatures, not from combustion processes in the cylinders. This lower ratio of pollutants holds true for the massive trailer trucks seen on the nation's highways despite the ominous-looking plume of black smoke that trails into the air from some of these vehicles.

Considering the low pollution levels and the very good fuel economy from a diesel, an observer might think it would pose a strong challenge to the passenger car field. (A typical Mercedes-Benz or Peugeot car diesel can provide up to 44 miles per gallon of fuel.) This challenge would seem to be strengthened by another advantage of the diesel. Its simplicity of operation and rugged design make it a very long-lived power plant. With regular maintenance, a diesel will continue to run well for years and years after a normal gasoline engine is ready for the scrap heap.

However, one reason why most passenger car drivers would not be overly impressed with a diesel is its lower performance compared to a gasoline design. Most diesel engines of equal size and weight, compared to a spark ignition engine, will have a lower top speed. A driver accustomed to hitting speeds 10 to 20 miles over posted limits on an open stretch of road feels tremendous frustration when his car is powered for only a few miles per hour over these restrictions.

The greatest minus for the diesel, however, is noise. Its clanking and banging make some drivers fear it will fall apart any moment. This noise is more noticeable in smaller engines because they must operate at higher engine speeds to provide performance close to that of gasoline engines.

There isn't much that can be done about the noise, though, because this is a part of the combustion process. Detonation—or knocking—is purely a function of burning. What the driver or onlooker hears when a diesel is running is the sound of the small explosions occurring each time a fuel-air mixture is ignited. This noise increases in intensity with the amount of compression needed before ignition occurs. Since the diesel is a very high compression system, it will automatically incur higher noise levels than lower compression engines.

It is harder to detect the engine noise in trucks or diesel-operated buses because these are medium speed engines. Also, the engine noise is dwarfed by the naturally high sound levels generated by the other parts of a very large vehicle.

Not only is the diesel unlikely to increase its hold in the passenger car field, it also must withstand competition from other kinds of engines such as advanced steam designs or gas turbines. Companies producing diesels counter this by continued study of improvements in diesel engine operation.

Among the areas in which advances have been reported are improved fuel-air mixing, use of special fuel additives and improved methods of supercharging.

All of these things are concerned with improving the combustion process so that there is higher power output. The limiting factor in power output for diesels (as well as spark ignition engines) is the amount of oxygen that can be introduced into the cylinder for each cycle. The greater the amount of air available in the cylinder, the greater the amount of fuel that can be burned per power stroke. Ways of increasing the oxygen content include increasing the revolutions per minute. The cylinder is then filled more often for a given period of time. Another way is to use a special pumping system to force more air into the cylinder for each cycle. (The second technique is called supercharging.)

In the first category, a potential breakthrough, developed some years ago by International Harvester, is based on introducing air into the chamber during the compression stroke with a rapid swirling motion. This swirl, reaching a circular movement as high as 20,000 revolutions per minute, results in faster and more thorough mixing of fuel and air. Thus, the compression-ignition process can be accomplished in less time than with a conventional diesel. Laboratory and road tests of the new engine in the 1960s indicated an improvement in fuel consumption to the point where only about half the fuel of a gasoline engine of the same power would be needed per cycle.

Fuel additives also have been valuable in improving engine power output as well as cutting down on the smoke from exhaust pipes. However, an even better method of reducing smoke was found to be the use of supercharging. Actually, this was an unexpected, additional gain from introducing supercharged systems on very large trucks.

Though supercharging has been used for other diesel applications for some time, its use on large trucks has become widespread only in recent years. What started the trend was a law in California that permitted a single-truck cab to pull two large trailers—each 40 feet long—instead of only one. After California gave its sanction, all the other western states followed suit. This gave truck operators the promise of much greater earning capacity because the same number of truck cabs could be used to haul twice as much cargo. The problem was that standard truck engines didn't have enough power output to haul the extra weight across the high mountain roads of the Rockies and other mountain chains.

The lower power output of truck engines relates to laws limiting their highway speed to about 55 miles per hour. It doesn't pay to have extra power for high speeds when the truck can't be used at them. So trucks have almost always used the smallest size engine for the weight requirements of the cargo. In general, a "large" truck diesel has a power output of only 250 hp., while some passenger cars may have engines capable of 350 or even 400 hp. This low power output is the reason trucks move so slowly on any kind of upgrade.

After studying the problem, engineers decided that supercharging could provide the extra efficiency and, therefore, the added power output for hilly operations. The higher pressure air compensated for the main problem in running engines at high altitudes—the thinner air found at heights of 2,000-5,000 feet or more above sea level. When advanced superchargers, such as those made by Schweitzer Corporation or the AiResearch Division of the Garrett Corporation, were installed, many new advantages resulted besides high altitude operating ability. For one thing, the turbulence that results from the pumping of high pressure air into the cylinders caused much better fuel-air mixing and reduced the fuel needed for a given power output. It was also discovered that the engine ran much more cleanly with an impressive reduction in smoke emission. As a result, supercharging is now in use throughout the country, even for trucks not scheduled to go through mountainous areas.

Work continues on improved supercharging methods, such as General Motors' RamAire system. This consists of an air compressor, storage tank, a valve, a pressure regulator and an ejector to introduce outside air into the system. The driver can cut in the supercharger by pushing the accelerator pedal all the way down. This trips a switch that opens the valve. This allows air from the tank, which has been pressurized by the belt-driven compressor to about 1,500 pounds per square inch, to go into the pressure regulator. The regulator reduces the pressure to about 150 pounds per square inch. The 150-psi. air moves the ejector

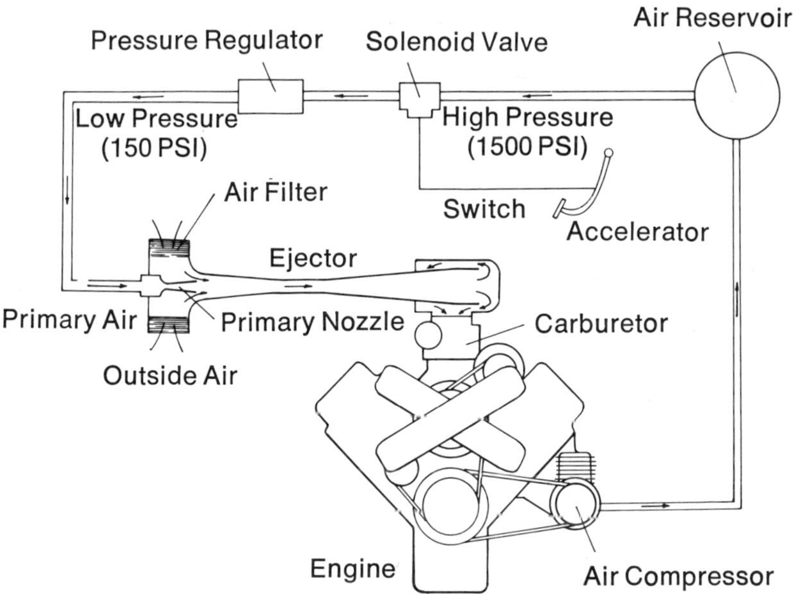

General Motors' RamAire supercharger system probably will be used in production diesels of the future.

where it draws in some outside air and continues on into the intake system. The action of the compressed air in inducing the flow of some cool outside air into the system, GM points out, permits use of a smaller storage tank than if the system used only stored air.

Supercharging as well as other improvements have increased the efficiency of the diesel and made it tougher for other engines to replace it. Most experts agree gas turbines will fight their way into at least part of the market in the future. But many believe it will be a long, long time before the diesel loses its hold on brute force needs unless it is eliminated by such things as new pollution requirements.

As one expert points out, the diesel has an almost unbelievable ability to continue running if taken care of properly.

"Turbines have costly alloys and, as of now, require very precise tolerances of bearings, shaft clearances, etc. These tolerances are much harder to maintain than a diesel, and contribute to the high cost of a turbine. A typical truck or bus diesel, on the other hand, goes 350,000-400,000 miles without major failures. The best proposed automotive turbine would be lucky to go 20% of this distance without needing rebuilding. And the diesel, after 400,000 miles, can be taken out, rebuilt for relatively little money, tested and put back in to go another 300,000-400,000 miles. Today, it's not unusual to find trucks, tractors and other diesel-run vehicles with over a million miles of service—and that's a record it's awfully hard to beat."

4

Jet Age Power—the Gas Turbine

While automotive engineers worked hard to solve the pollution problems of the internal combustion engine, other kinds of engines were being studied in case these efforts failed. The other engines fell into two main categories: engines that did not use fossil fuels at all, such as electric designs, and those that still used such fuels but burned them continuously rather than intermittently as did the internal combustion engine.

The continuous combustion candidates included steam engines, Stirling-electric hybrids and gas turbines. Of these three, the type that seemed most likely to take over an important share of the automotive power market by the late 1960s was the gas turbine.

The gas turbine had a number of potential advantages

from the standpoint of performance. It started out with improved anti-pollution properties compared to the conventional automobile engine without pollution control devices. Equally important, the gas turbine industry was not in its infancy like the hybrids or other "exotic" engines. Production of gas turbines, in fact, rivaled in dollar volume the production of internal combustion engines, though these turbines were for use in airplanes or in industry. Starting in the years after World War II, gas turbines had almost completely ousted piston engines from their previous domination of the aircraft field. From the 1960s on, they had started moving into the electric utility field. The use of gas turbines to run electric generating plants has steadily increased at the expense of such other units as steam turbines.

Under these conditions, the gas turbine field had built up a very respectable amount of technical experience. Efficient turbine parts had been developed for plane and industrial usage. A great deal of basic information on the combustion processes and the gas flow patterns was obtained.

This did not mean that the turbines for other uses could simply be modified and installed in working automobiles. Actually, no matter what basic type of engine is considered, the details of an engine will vary greatly from one purpose to another. However, having a good understanding of basic principles is always the first vital requirement in taking a new system from the sample stage to practical production status.

A review of the basic principles of the gas turbine, as well as the example of its steady inroads into the aircraft market, was enough to encourage many automotive companies to start studying it as early as the 1940s. In the 1940s and '50s, of course, there was little thought given to the anti-pollution advantages of the gas turbine. Automotive engineers became interested in it because it promised some important gains in performance. These gains were not considered to have major value for passenger cars. Rather they seemed better suited for heavy duty applications, such as trucks, buses, heavy construction equipment. Indeed, the first important automotive uses of gas turbines came about in those areas in the early 1970s.

Major basic advantages of turbines for automotive use include good "torque" properties; multi-fuel operation; light weight; relative simplicity; smooth operation with minimum vibration; and all-weather operation. The following short discussion shows what all this means:

Torque is the force that produces rotation. Torque properties refer to the amount of rotational energy the engine puts out. (Torque, a technical term, equals the product of the force applied to a lever (or shaft) multiplied by the distance to the point about which the force rotates.) In the case of the gas turbine, it takes lower shaft rotating speeds to turn out a higher amount of torque than other engines.

The second advantage mentioned is that a gas turbine can work with many different kinds of fuels. Though the efficiency may not be the same from one fuel to the next, the same engine can run on all kinds of diesel fuels, jet engine fuels, and kerosene. The turbine also will work with almost any kind of regular gasoline, but its operation is drastically harmed by lead. The lead tends to coat some of the working parts and change their operating properties. If too much lead coats these parts, the engine will stop operating. However, the new restrictions against the use of lead in standard gasoline mean that unleaded gasoline is rapidly becoming available at all gas stations.

The light weight of the turbine is related to its simplicity as well as its high power output per pound of fuel. The engine doesn't need the pistons, cams, forged crankshaft and flywheel parts of the reciprocating engine. In addition, it doesn't require liquid cooling equipment nor does it use up oil, because the oil is separated from the working areas. As an example of what this means, the first production truck turbine developed by General Motors' Detroit Diesel Division (designated GT-309) weighed only half as much as a standard truck diesel engine of equal power.

The smoothness of operation of the turbine comes from the elimination of pistons and of the intermittent running properties of internal combustion engines. The turbine is always rotating when the engine is in operation. Thus there are no sudden split-second stops and starts as in the internal combustion engine. The smooth operation of the turbine also has another advantage—it insures lower noise levels than piston type designs. Some people might question this, considering the thunderous noise of a huge jet plane taking off. However, the power in these engines dwarfs that of the most powerful car on the road today. The total output of the four engines of a 707 or DC-8, for instance, is the equal of hundreds of high performance cars.

The all-weather operation advantage means that the operator of a vehicle using a gas turbine doesn't have to worry about using special coolants for hot weather running or anti-freeze for even Alaskan winters. A turbine will start and warm up rapidly under the most extreme temperature conditions.

At the same time, the turbine still had its drawbacks for passenger car use as the 1970s got under way. For one thing, there is a greater time lag for movement of the car's controls to reaction of the vehicle at lower speeds. This isn't too bothersome for long distance, freeway-type driv-

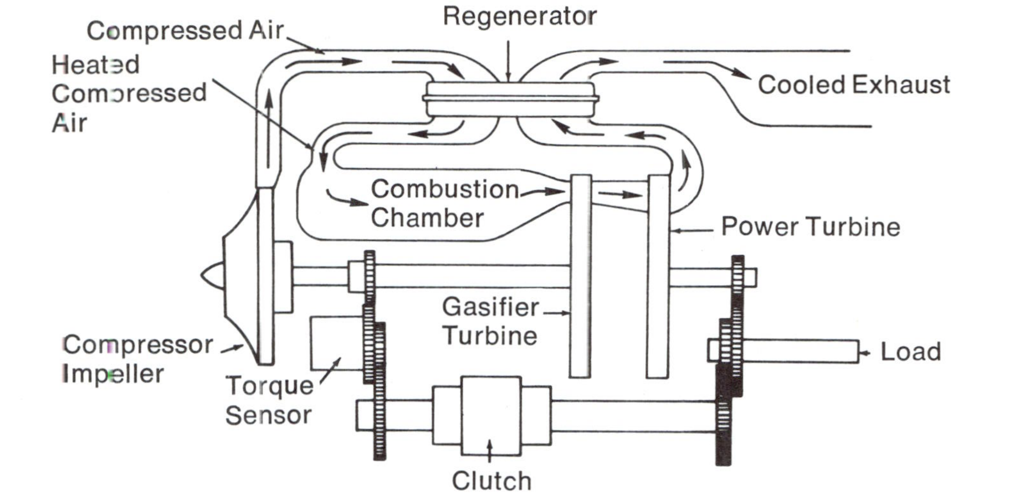

280 HP TURBINE SPECIFICATIONS

Rated Power	280 hp	Rated Gasifier Speed	36,000 rpm
Rated Torque	575 lb. ft. at 2560 rpm	Rated Airflow	3.9 lb./sec.
Max. Torque	1150 lb. ft. at 0 rpm (stalled output)	Rated Compressor Pressure Ratio	3.8:1
Max. Speed	2850 rpm	Engine Weight	1425 lb. (approx.)
Rated Turbine Inlet Temp.	1700°F	Engine Dimensions (L x W x H)	45 x 26 x 40"

Operating diagram of the 280-hp gas turbine engine for trucks developed for General Motors by its Detroit Diesel Division.

ing, but is hard for a driver accustomed to stop-and-go city driving. For drivers of trucks, buses and other bulky vehicles, of course, a certain amount of lag before their massive carriers get moving is a normal part of the job. Solving such control problems for smaller cars was not impossible or even extremely difficult. It just required more years of engineering studies. The fact that gas turbines would raise the price of cars considerably until true mass production designs were in hand was a greater deterrent to their immediate introduction in the passenger field.

At any rate, by the mid 1950s, almost every company in the automotive field was building and testing gas turbines. The big three of the passenger car market, Ford, Chrysler and General Motors, all had programs going, including some studies of passenger types. Chrysler, in particular, concentrated a great deal of effort on family car possibilities. The company developed a workable 140-horsepower model and demonstrated it in a 1954 Plymouth. After further improvements, a 50-car test program got under way in 1965. For several years, the cars were loaned to a specially selected cross section of drivers in different parts of the country for periods of several months each. The data from this program was still being evaluated at the start of the 1970s. The Chrysler turbine weighed about 450 pounds, had a torque of 375 foot-pounds at zero revolutions per minute, and took up only about half the space of the normal engine for these model cars.

The General Motors Research Laboratories also ran small turbines in their special Firebird test cars. This program began in 1953 and was continued through many advanced models of Firebirds, but all of these were one-of-a-kind designs. The large scale turbine work at General Motors, as at Ford and other firms, was on large, heavy duty truck types, such as the GM GT-309, which turns out almost 300 hp. at 3,600 rpm., or Ford's 707, which developed 600 hp. at 3,080 rpm. Much of the development of turbines has been done by companies concerned only with heavy duty equipment, such as Mack Trucks, International Harvester and Caterpillar Tractor.

The basic principle of the gas turbine is not only simple, but it is very old. The first working model of a jet engine goes back over 2,000 years. Roughly 100 years B.C., the scientist Hero, living in Alexandria, Egypt, invented an engine he called an *aeolipile*. He suspended a hollow sphere with two small pipes sticking out of it on tubes that ran from the sides of the sphere into a kettle. The kettle, filled with water, was placed on a tripod over a wood fire. When the fire made the water come to a boil, the steam ran through the tubes into the sphere and out through the two small nozzles extending from it. The result was a rapid rotation of the ball.

Hero's tests were interesting, but he had no real idea of putting them to any practical use. Nor did anyone else for a long time to come. In fact, it took well over 1,600 years before there was any understanding of why Hero's machine worked as it did. It was not until almost the end of the 1600s that the great English scientist, Isaac Newton, discovered the Laws of Motion. These principles laid the groundwork for almost every major technological advance man has made since then, from the steam locomotive to the moon rocket. Newton's second and third laws explained the *aeolipile* and, more than a hundred years later, the motion of a balloon.

The second law states that the force exerted by an object is equal to the mass of that object times the rate at which it accelerates.

The third law states that every action has an equal and opposite reaction. In other words, the thing that causes an

Executives of Mack Truck Corporation examine the installation of an experimental gas turbine engine in a standard Mack-F cab-over-engine truck chassis.

action, such as a mass accelerated in one direction, is itself sent in the other direction by a force equal to the one originally created.

Taking Hero's system, if the force pressing against the inside of the sphere were generated by something inside the sphere itself, and the sphere was free to move, it might fly through the air. This is what happens with a toy balloon when it is pumped up with high pressure air and then released with its narrow end open. As long as the inside pressure in the balloon (or the sphere) is greater than the outside pressure, the balloon will move in accordance with Newton's laws. In this case, the action—the force generated by the escaping high pressure air—causes an equal reaction that pushes the balloon in the opposite direction.

This is the way a gas turbine works. The problem was to develop methods of continually generating the high pressure air so that motion doesn't stop as long as the operator wants the vehicle to keep going. The problem was far from simple, as is amply demonstrated by the 2,000-year time lag between Hero's engine and the first successful jet engines. However, in the years just before World War II, British engineers succeeded in developing working turbines that found their first use in supplying electrical power for missiles and rockets. Later, more powerful turbines were designed that rapidly became the main power sources for both military and commercial planes.

As was noted earlier, these successes caused many experts to start considering the use of turbines in cars. However, there were a great many new obstacles to overcome in order to do this. For instance, turbine engines normally operate best at full power. This isn't a great problem for planes, where a lot of power is needed, but for automobiles, where much of the operation is at low or intermediate speeds rather than all-out high speed freeway driving, it poses problems of developing an efficient engine.

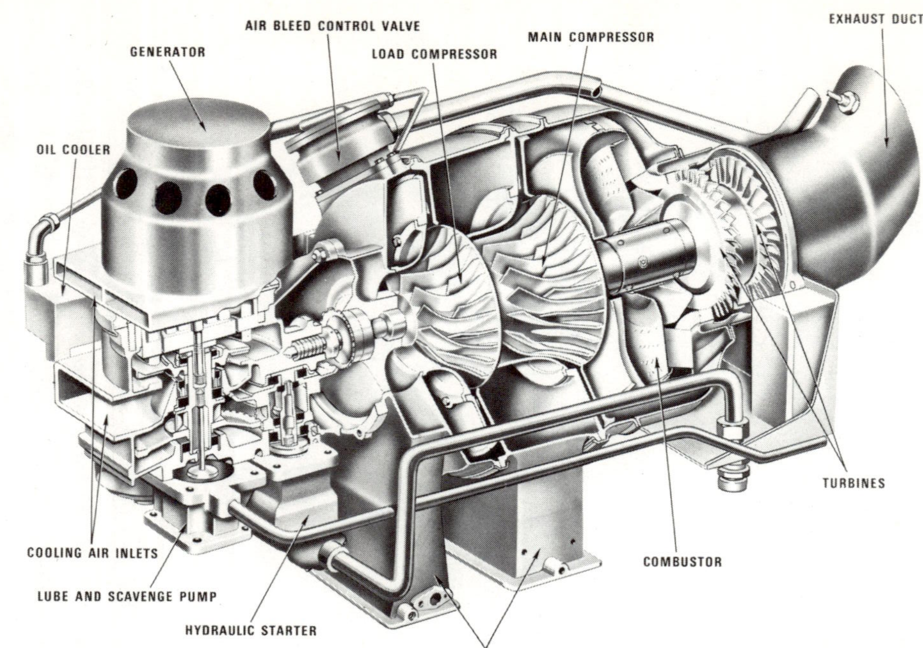

Cutaway view of a small turbine engine developed as an engine starter for military planes shows the simplicity of this kind of power plant. A major difference between this engine, developed by Williams Research Corporation, and an automobile turbine is the absence of regenerator disks in the aircraft design.

An even greater difficulty in modifying the turbine for automobile use is the temperatures inside the engine. It's hard to cool such key parts of the turbines as the turbine blades and injection nozzles. Therefore, these must normally run at temperatures almost as high as the hot gases flowing through the engine. As a result, in airplane jets, these parts are made from special, costly, high temperature materials such as titanium or special steels called superalloys. For giant airplanes, even costs of $400,000 or $500,000 for an engine aren't out of line with the total plane cost. In addition, if we're talking about a commercial airliner, the plane may cost a lot, but it brings in a lot of money on each trip because of the great many paying passengers it can carry.

Still another problem for a car turbine is finding ways of reducing the great heat output from the high temperature exhaust gases. On planes, much of this heat is used for such things as cabin air conditioning or running other systems. In addition, there's a lot more room for cooling air circulation from the surrounding air. In a car, some way had to be found to keep gases at 1,500 to 2,000 degrees from exhausting into the air.

It took time, but methods have been developed to meet most of these automobile drawbacks. Some of these solutions will be covered during a discussion of the general makeup of today's automotive turbines.

The basic parts of any gas turbine include an air intake, compressor section, a heat exchanger, combustion chamber and turbine section. The compressor and turbine sections contain rotating wheels with scoop-shaped blades extending from the wheel hub. Each one of these wheels is referred to as a "stage." A plane engine has many stages in both the compression and turbine sections. Examination of today's automobile turbines shows that almost all of them use only three turbine wheels—one in the compressor section and two in the turbine area.

This great reduction in stages and careful design of low cost ways of assembling and mass producing them is one approach to getting engine costs down. When the engine is running, the first important operation is scooping of air into the compressor section through the intakes. The single whirling compressor wheel is shaped to force the particles of air into a much smaller volume. In an engine like that designed by Williams Research Corporation, for instance, the air flowing from the exit duct of the compressor section into the engine has been pushed together so it has a pressure four times that of atmospheric pressure.

The compression step serves to raise the air pressure to about 370 F. This is well below the temperature needed for ignition in the combustion chamber, so an additional preheating step takes place before the air enters the chamber. The compressed air passes through a rotating, very hot disk called a regenerator. (There are typically two intake ducts and two regenerators, one on each side of the engine.) The regenerator gives more heat to the air, raising its temperature to about 1,150 F.

Before following the progress of the compressed air through the engine, we might examine the way the regenerator works and its importance in modifying the gas turbine for auto use. The regenerator disk is basically a rotating heat exchanger. A heat exchanger is a device that transfers heat energy from one place to another. In a piston engine car, for instance, the car radiator is a form of heat exchanger. The invention of the regenerator not only provided a way of preheating the intake air to increase efficiency, it also solved the hot exhaust gas problem.

As the regenerator rotates, its specially designed surfaces pass through both the compressed air intake ducts and the

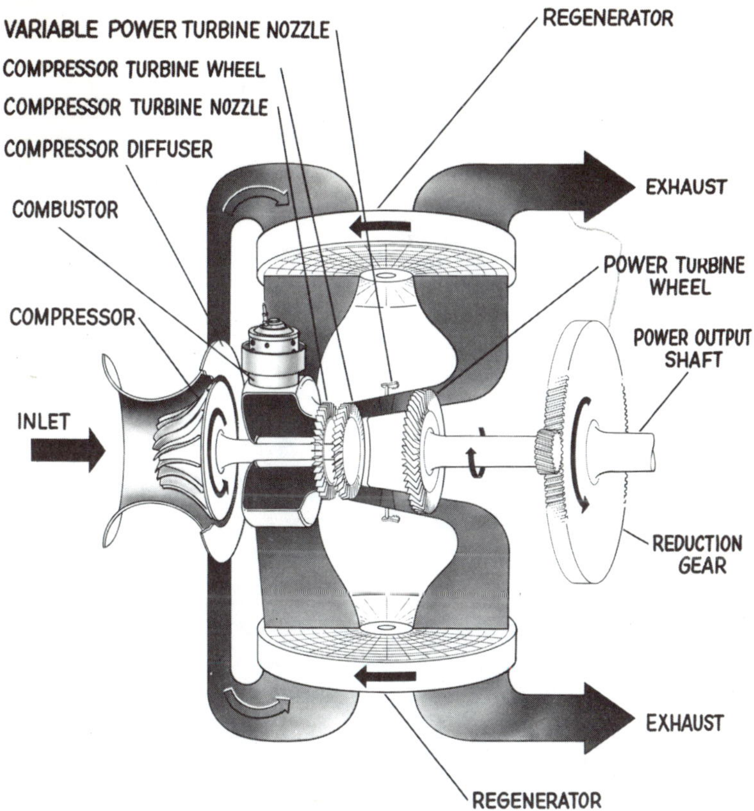

Schematic of the air flow through the Ford gas turbine engine shows the positions of the compressor and three turbine wheels as well as the action of the regenerator disks. Note that no mechanical connection exists between the compressor turbine wheels and the power turbine wheel.

exhaust gas ducts. In this way, it picks up much of the heat from the exhaust gases so that the temperature of the exhaust coming out the tailpipe is low enough for safe operation. The heat from the exhaust is then carried around by the rotating disk and transferred to the incoming compressed air. This action not only heats the incoming air, it also relieves the disk of its heat load so that it is ready to pick up new heat energy from the exhaust.

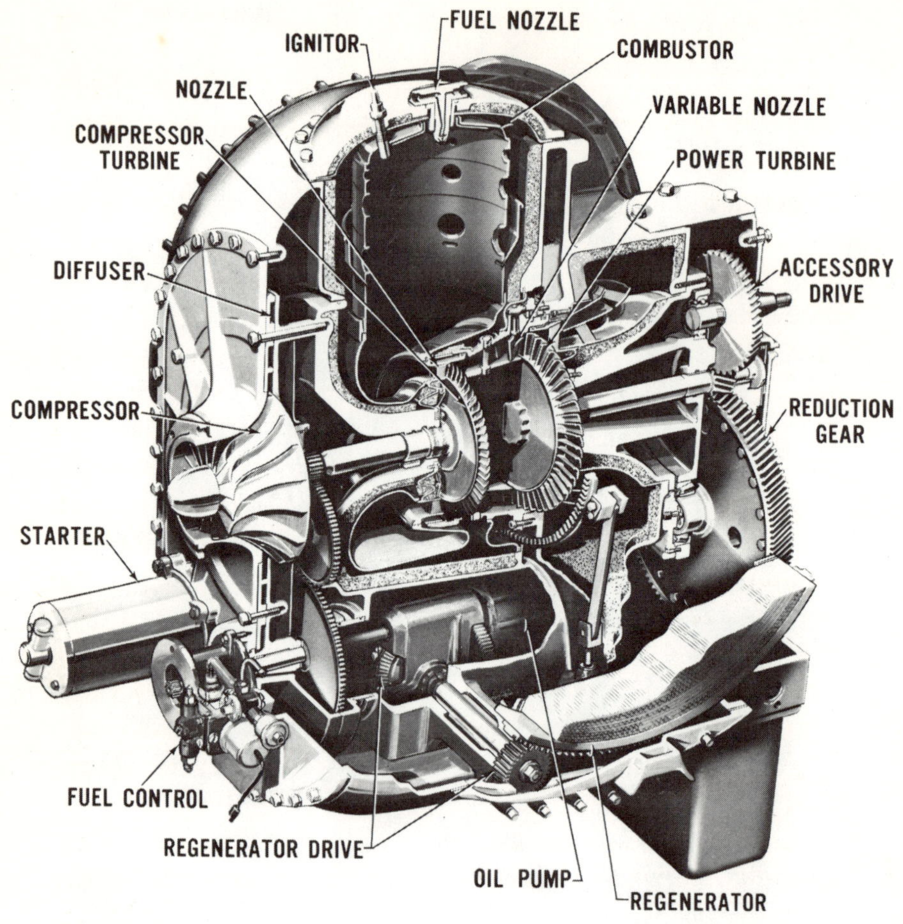

Cross section through the production version of Ford Motor Company's large gas turbine engine.

Going back to the compressed air, after passing through the regenerator, it enters the combustion chamber. Here fuel is sprayed into the chamber to mix with the air. There are a number of ways of inserting the fuel. In the case of the Williams engine, a rotating fuel inlet with many tiny holes in it is installed on the mid-section of the shaft connecting the compressor and the first of the two turbine stages (called the compressor turbine). The Chrysler engine first used in its passenger car tests of the 1960s had a separate series of fuel nozzles leading into the sides of the combustion chamber for this job.

In the combustion chamber, the fuel-air mixture is ignited, raising the gas temperature to about 1,700 degrees F. in the designs of the early 1970s. (The significance of varying this temperature will be shown a little later.) The hot gases coming from the combustion chamber expand, giving up much of this energy by striking the blades of the compressor turbine, causing it to rotate. This action provides the rotating motion that runs the initial compressor stage.

Usually the second turbine stage is larger than the first. This is because the second stage, called the power turbine, is the one that transmits the rotating motion to turn the vehicle's wheels. The hot gases, still retaining most of their energy, flow past the compressor turbine to strike the blades of the power turbine. After spinning the power turbine, the gases, still at about 1,300 degrees F., move into the exhaust duct line that takes them through the regenerators.

A major need in the turbine section is some way of "unhooking" the power turbine from the compressor, either partially or completely. On a plane, this isn't needed since most of the flight takes place at a fairly constant speed. There are no traffic lights or cross traffic in the sky. A car or truck, obviously, needs to be able to make a great many changes in speed—up or down. This is accomplished by varying the amount of gas that can strike the second-stage turbine blades.

The power turbine is thus not attached to the same shaft as the compressor turbine. In the case of the Chrysler engine, there is no direct connection at all between the two turbines. Instead, there is a small gap from one to the other. In the Williams design, the power turbine shaft passes inside the larger compressor shaft and is supported on a special rotating bearing.

With this arrangement, it is possible to uncouple the power turbine shaft when the vehicle is not moving, while the engine continues to run. This is done in almost all of today's turbines by using movable vanes in the ducts between the two turbine stages. These vanes can be turned so that all the gas flow from the compressor goes around the power turbine directly into the exhaust. Or the vanes can be positioned to allow as much gas as desired to hit the power turbine vanes, turning the turbine at faster and faster or slower and slower rates. In general, the gas flow control vanes are connected by linkages to the driver's controls so that they move in or out of the flow depending on the position of the accelerator pedal.

A great amount of research effort is going into studies of obtaining higher gas temperatures in the combustion chamber. There is a direct relationship between the temperature of the gases flowing past the turbines and engine efficiency. The higher the temperature, the more energy the gases will have to impart to the blades, and the faster the turbines will spin. The main problem is finding materials that will take temperatures above 1,700 degrees safely for long periods of time. If this is done so that a temperature of, say, 2,100 degrees can be gained, the same size engine could turn out 40% more power. Conversely, it could turn out the same power as at 1,700 degrees for 40% less engine weight.

The gas turbine, by the start of the 1970s, had become practical enough for use in heavy duty vehicles. Though it still cost more initially than some competing piston types, bus companies, truck firms and construction users could see that lower maintenance costs, higher efficiency and other advantages made the turbine very attractive for years

Test beds for 335-hp Ford turbines in the early 1970s included a Continental Trailways bus and a large trailer truck.

As this artist's sketch indicates, turbine engines may make Dick Tracy-like flying platforms a common vehicle of the future. Combinations of turbines and wheeled undercarriages may permit cars that can be efficiently converted to planes.

of operation. As a result, Ford, General Motors Detroit Diesel Division and several heavy duty vehicle firms announced availability of high-powered commercial turbines. The gas turbine industry trade magazine, *Gas Turbine World*, estimated in 1971 that the market for heavy duty engines (diesel type) would go from 160,000 units a year to 320,000 a year by 1980. While turbines were not challenging the diesels in 1971, *Gas Turbine World* predicted that, by 1980, most of the 320,000 engines sold would be gas turbines.

The passenger car market still seemed some years off in the early 1970s. However, promising lightweight engines were being built, and the U.S. Government provided funds for a development program for engines to replace piston types if the latter couldn't overcome pollution problems by the mid 1970s. Thus a five-year program for both gas turbines and steam engines was started in 1971.

The timetable established by the U.S. National Air Pollution Administration called for completion, by early 1973, of four different prototype small turbines rated from 80 to 200 hp. While this went on, NAPA funded special research and development programs on the most critical parts needed for a practical, low cost family car turbine: low emission systems, improved regenerators and high temperature material systems. The goal was to have five test cars running in 1975, leading to possibly larger scale fleet tests for the last half of the decade.

As early as 1971, there was evidence that the small turbine could meet the performance goals. A particularly important example was the series of tests started in New York City on a design prepared by the Williams Research Corporation, Walled Lake, Mich. This engine was developed by an engineering group directed by Sam Williams, a key member of the team that had designed the engines used in the Chrysler Corporation passenger car tests of the sixties.

Sam Williams, president of Williams Research Corporation, stands beside his firm's low pollutant emission gas turbine engine, completed for tests in the early 1970s.

The basic engine had an output of 80 brake hp. at 4,450 rpm. output shaft speed and torque of 180 ft.-lbs. at zero output shaft speed. Basic engine dimensions show why it fitted so easily into a compact car: length, 24″; width, 26″ and height, 16″.

So the gas turbine, after several false starts in the 1960s, was on its way. For car use, though, it still has to win out over both the old reliable piston types and quite a few other "exotic" proposals, from steam to hybrid designs.

5

Steam Engine Comeback

Ames, Artzberger, Binney-Burnham, Brecht, Crouch, Gardner-Serpolet, Holyoke, Knoxmobile, Pawtucket, Remel-Vincent, Tractmobile—strange-sounding names all! What do they have in common? All were once production-model steam engine automobiles available to the U.S. public. Not only that, they are only a small sampling of the 141 different makes of steam cars and trucks produced in this country between the late 1890s and 1020.

In the very early days of the "horseless carriage," in fact, the race for domination of the automobile power market seemed to lie between the steam engine and the electric power plant. In stating its case for steam, Stanley Motor Carriage Company, maker of the most renowned family of these cars, the Stanley Steamers, advertised in the early decades of the century that this was "A Power Plant Stan-

73

dardized One Hundred Years Ago . . . George Stephenson laid down the principle in 1820. And the principle he definitely fixed is that of a two-cylinder engine, double acting (taking steam in each direction), simple (using the steam in one expansion only), with slide valves and link motion reverse."

When automobile buffs of the early 1900s got excited about new speed records, it was usually a steamer that set them. In 1906, for instance, Fred Marriott set the world land speed record in a Stanley Steamer. He didn't dodder along at a speed only slightly faster than a horse, either. He hit a mark of 127.66 miles per hour over the sands of Ormond Beach, Fla. The following year, he achieved a speed of 190 mph. in a steamer before the engine broke down. Such speeds could only be maintained over short distances, but the most finely tuned stock cars on the race tracks of the 1970s will only average 140 to 150 mph. in most cases.

At the height of its popularity, in 1916, the typical steamer family car had a respectable cruise speed for those times of about 30 mph. Such a design was the Stanley roadster, which was a two-cylinder, double-acting model rated at 20 hp. and with a range of up to 160 miles. The engine had only 13 parts and, because of this simplicity, was one of the most reliable on the road.

Despite all this, the steamer was fighting a losing battle in the World War I years against the internal combustion engine. The latter lent itself to mass production methods and could be turned out at much lower cost. Not only that, but the internal explosion of the fuel-air mixture could generate much higher energy levels, which translated into faster and faster average speeds, for lower weight. With the limited knowledge of those years, steam engine designers were faced with the problem of having to build bigger and bigger steam boilers to increase the power output that could match the I.C.E. in performance. For these and other competitive reasons, the steam engine lost the battle. By the start of the 1930s, it was considered to have joined the stagecoach as a transportation system of the past. From 1929 to the 1970s, no new production models of steam cars appeared on the market. And, for most of this period, little or no research on steam car engines was performed.

A few diehards did continue to look into updating steam systems. A man named Fred Besler acquired patent and design rights from one of the last steam firms, Doble-Detroit Motor Company. He continued to modify and improve the system during the 1930s and '40s. In the 1950s, a group at Studebaker Motor Co. looked into possible steam power plants. Occasional research studies also were made at the major U.S. car makers' laboratories. None of this work breathed new life into the steam engine and it seemed as dead in the mid 1960s as it had seemed in the mid 1920s.

In the late 1960s, however, the furor over smog arose, and suddenly groups in many parts of the country were re-examining steam technology. Environmentalists, both inside and outside government, noted that steam, at least, offered one way of drastically cutting auto engine pollutant emissions.

Someone unfamiliar with the way a steam engine works might conclude it has nothing to do with fossil fuel. This, of course, is not the case. The steam engine requires fuel to supply heat to turn the working fluid—which can be water or some other suitable liquid—into vapor form. However, unlike the internal combustion engine, this fuel isn't restricted to the relatively small volume of a cylinder chamber for the burning process. As a result, the burner system can be designed to take in all the air that is needed for full combustion of the required amount of fuel. The burned gases from this system, then, are much lower in pollutants than is an internal combustion engine's exhaust.

Aiding in this is the fact that this kind of engine can use many different kinds of fuel. It can operate using any standard grades of fuel available at any present day gas station. Its performance, though, will be just as good whether the fuel is one of the new unleaded gasolines or the high octane mixtures required for the internal combustion engine of the 1960s and early 1970s. It can also work just as well on low grade white kerosene or gasoline that is not as highly refined as the products of the early 1970s. Thus a vapor-cycle engine gives the promise of operating with a fuel that is not only less polluting, but also less expensive. In addition, tests run by a number of experimenters have indicated that a vapor-cycle engine can give gas mileage on low cost fuel as good as that provided by internal combustion engines using high octane fuels.

The basic principles of all vapor-cycle engines are the same, though the system details can change quite a lot from one design to another. The heart of this engine is the conversion of a liquid substance to its vapor state so that the energy stored in the vapor can then be released to drive the mechanisms that turn the car's wheels. There are many kinds of fluids other than water that can be used, with various advantages and disadvantages compared to the water-steam cycle. This is why the term "vapor-cycle engine" is a more appropriate description of this class of power plants than steam engine. (The energy cycle the working fluid follows in these kinds of engines also has a more technical name—Rankine cycle, after the English physicist who first described it—just as the internal combustion engine works on the "Otto" cycle.)

The first step in any vapor-cycle engine is to ignite the fuel in the burner in order to heat the working fluid which is circulating through a series of tubes around the combustion chamber. The combustion system is designed to provide a high enough temperature to change the liquid into a vapor. A series of valves controls the passage of the vapor so that it is first raised in pressure and then allowed to expand at the proper rate against a device that, in turn, translates the energy from the vapor into rotation of the shaft. Different vapor-cycle engines use different devices—some use reciprocating pistons, others may use a turbine, while still other types have been studied with rotating pistons.

In most of the old-time steamers, the vapor, after it had done its job in moving the drive mechanism, was simply exhausted to the air. With this kind of "open" Rankine cycle, large storage tanks were needed to hold enough fluid for the car to travel a reasonable distance before refilling was required. Compact systems had not yet been developed which could condense the vapor back to liquid and use it again. In the decades after the last steamers of the 1920s were produced, considerable new knowledge has been gained in development of closed systems for such uses as home air conditioning or for manned space flight.

As a result, today's new vapor-cycle engines use the "closed" Rankine approach. The exhausted vapor is led into a device called a condenser. This looks much like the radiator on a standard internal combustion engine. Here the vapor is cooled so that it returns to its liquid state. Then it feeds right back into the coils around the combustion chamber for reheating and reuse in running the engine. With the sealing material and methods of today, engineers who work in vapor-cycle engine development state the loss of fluid is so small that a car should be able to go for at least a year, possibly a number of years, without needing new fluid.

One of the major advantages of a Rankine type engine is

that it delivers high torque at low speeds. As a result, the engine can be used without a transmission if desired. In fact, the car can accelerate from a dead stop without any system of gears or hydraulic cylinders to step up the engine output for starting. In studying different kinds of engines in its search for a pollution-free power plant, the U.S. Senate Commerce Committee reviewed starting properties of a steam and an internal combustion engine. Said the report: "An engine (steam) with an engine rating of 168 horsepower will accelerate from 0 to 60 miles per hour in 9.5 seconds. An I.C.E. with a rating of 250 horsepower propelling the same load takes 10 seconds to accelerate from 0 to 60 miles per hour."

The reason why an I.C.E. needs a transmission while a vapor-cycle engine can get along without one lies in their speed properties. The I.C.E. is basically a constant speed engine. Though its speed can be changed by changing the amount of fuel fed to it, the engine will run inefficiently at all but one optimum speed for a particular engine design. The best way to allow the engine to run at or near its best speed all the time, while still letting the car move along at varying speeds, is to change the power sent to the car's wheels. Such things as step-up or step-down gear systems are used. In other words, the engine runs about the same rpm., but the transmission changes the ratio of engine-operating power to the rotating rate of the four tires. The vapor engine, on the other hand, will run efficiently over a wide range of vapor pressure. These pressure changes are made by moving valves in or out to meter the vapor flow into the cylinders or against the turbine blades.

However, engineers have found that it can be worthwhile to include a transmission in a vapor-cycle engine. The transmission is used not so much to regulate the power sent to the wheels as to tap some of the energy to

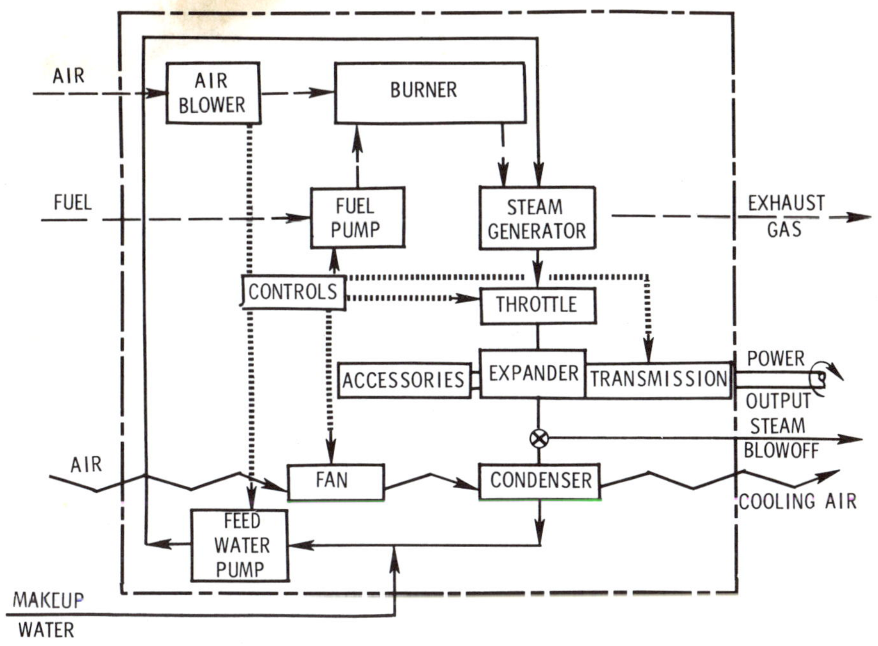

This simplified block diagram indicates the way the SE-101 steam engine operates.

operate other car systems—air conditioning, radio, etc. Some of the proposed vapor-cycle transmissions use novel methods, as an examination of the General Motors SE-101 will demonstrate.

The SE-101 was one of two vapor-cycle engines being tested in GM production-type cars in the early 1970s. The second, called the SE-124, was developed by the company founded by Fred Besler. The firm, Besler Developments, Inc., of Oakland, Calif., provided an advanced Besler model engine under contract to GM. At the start of the 1970s, GM had gained a great deal of data on vapor-cycle systems from using the SE-101 in a modified Pontiac Grand Prix, and the SE-124 in a Chevelle sedan.

The method of starting the SE-101 (and the SE-124) seems no different than any conventional car. The driver

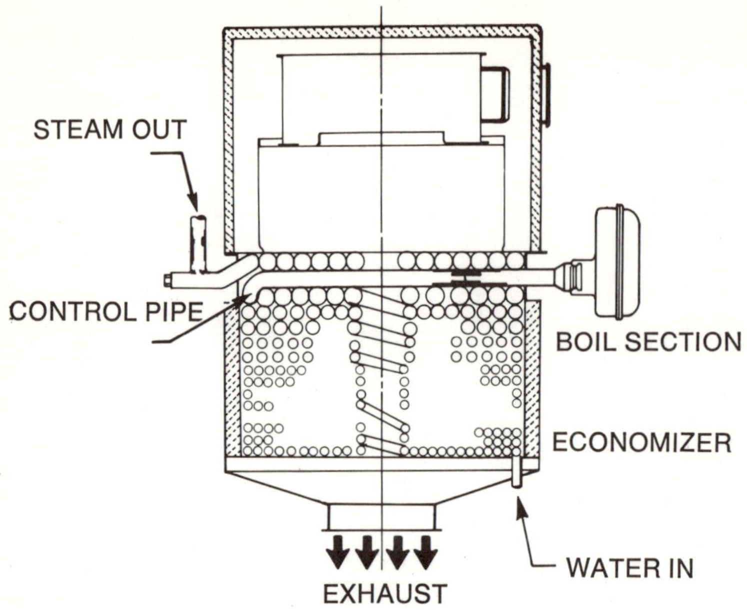

Installation side view of the Besler SE-124 steam engine highlights the use of the two large cylindrical sections in the steam generator-combustion chamber operation.

turns the ignition key to *on*, waits about half a minute until a dashboard light indicates the proper steam pressure has built up, then pushes down on the accelerator and moves off. What is going on under the hood is, of course, not the same as in the passenger cars using internal combustion engines.

The first thing that happens when the key is turned is that an electric pump fills the boiler with water. When the right water level is reached, it touches a sensing unit that sends an electric signal to start an electric motor. The motor is used only to make the combustion blower and fuel pump start running. Once these have begun the combustion cycle, this motor shuts off until it is needed for another start.

The blower is used to suck in air through the air intakes

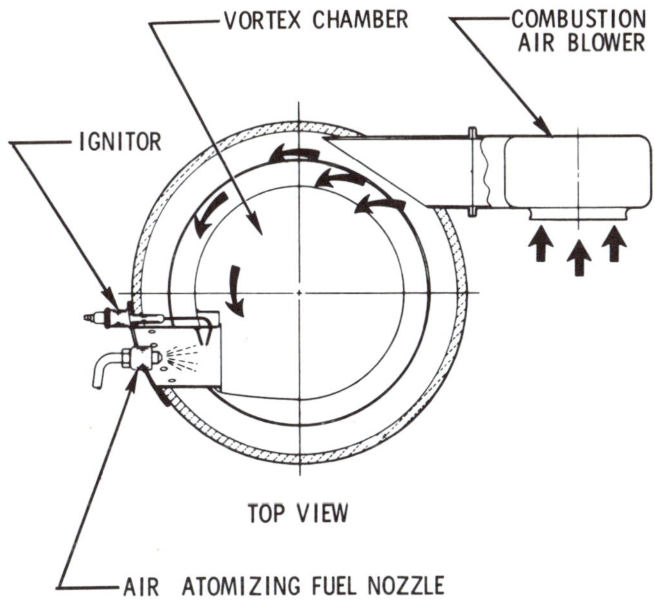

Top view of the SE-124 combustor shows the swirling movement imparted to the air coming into the chamber.

and blow it into the combustion chamber. At the same time, the fuel pump extracts a small amount of fuel from the fuel tank and begins to continuously spray millions of tiny droplets of this substance into the chamber. In the SE-101, the mixing of fuel and air takes place in two chambers containing small rotating turbine wheels. A spark plug ignites the fuel-air mixture and the resulting hot gases flow over the engine's steam generator. The generator consists of several sets of small-diameter steel and stainless steel tubes which are connected and folded close to each other in a staggered arrangement. If the tubes were stretched out into one single straight tube, they would cover a distance of 430 feet—half the length of a typical city block. Arranged as they are in coiled form, the SE-101 is only 31 inches wide by 15.5 inches high by about 12 inches deep.

The high pressure steam from the generator passes through a throttling valve into a crank-and-piston expander. This is an arrangement similar to that used in an internal combustion engine. That is, the power section consists of four cylinders, each containing a piston linked from the bottom end to a crankshaft. When the engine is running, a camshaft is used to open the cylinder inlet valves briefly for each cylinder in order. As one of these valves pops open, a certain amount of high pressure steam rushes into the cylinder. This pushes the piston downward, thus causing the crankshaft to rotate a certain number of degrees. As each of the cylinders goes through this step, the crankshaft is made to rotate rapidly as long as steam is applied to the pistons.

In each expander cylinder, the steam loses its energy as it pushes the piston down until, at the bottom of the stroke, a series of openings in the cylinder wall is uncovered. The now low pressure vapor passes out through these ports into the return lines to the condenser. In the case of the SE-101, the inside diameter of each cylinder is 3.5 inches and the overall stroke is 2.625 inches.

The SE-101 condenser consists of two connected heat exchangers. These exchangers are made up of many loops of tubing to which are attached large surfaces of heat-conducting metal called fins. The material used for tubes and fins is aluminum. Changing the working fluid from a vapor to a liquid calls for lowering the fluid temperature to a level where condensation takes place. In the SE-101, this is accomplished by using two high speed fans to pass cooling air over the heat exchangers. The fans, which are typical automotive-type designs, require 18 horsepower to operate. The air driven past the aluminum surfaces picks up heat sent into the aluminum by the fluid and takes this excess heat into the surrounding atmosphere. By the time the fluid has gone through the long distances inside the exchangers, it has been cooled down until it comes out into the feedwater system as a liquid again.

Going back to the crankshaft operation, the rotating energy from this device is directed to the car's wheels through an unusual transmission called a toric transmission. This system receives its name from the mathematical shape called a toroid, which is applied to its workings. It was used in the SE-101 because tests showed it promised a savings of 17% over a conventional transmission in steam needed for a given horsepower output. The SE-101 tests also permitted study of a transmission method that may eventually find uses in many of the cars selected for future vehicles, including gas turbines.

The toric section consists of two discs, called races, which are machined to form a toroidal space. Three rollers are inserted between the races, and these rollers are mounted on carriers which are fixed to the case. By tilting the rollers in relationship to the axes of the races, the speed ratio can be changed. That is, the speed of the output race can be changed compared to the speed of the input race. Power is transferred from one race to the other by traction in the places where the races are squeezed against the rollers. The change in position of the races is made by an automatic hydraulic system that senses changes in acceleration based on the driver's pressure on the pedal.

Other parts of the system, including the fluid torque converter, transmit the rotations per minute from the output race to the wheel system. Systems that take some of the power sent into the transmission and use it to run the car accessories are an important part of the transmission network. If there were no transmission system to tap for this, the designers would have to add separate electric motors or other power devices to run accessory systems.

The basic steps in the SE-124 Besler engine are the same as in the SE-101, but there are quite a few changes in the details. As in the first case, a burner system is used to provide hot gases to turn the working fluid into a vapor. The vapor, however, is not expanding once, as in the SE-101, but twice, to use more of the energy in the vapor and increase engine efficiency. This is done by first directing the high pressure steam into a small, high pressure cylinder (2.5 inches inside diameter). Then, after it moves the piston over the desired stroke, it is sent to a larger, low pressure expander (4.25 inches inside diameter) to again move a piston. The expander system, unlike the SE-101, is double acting. That is, the steam is alternately ducted into one end of the cylinder, then to the other, moving the piston back and forth like a shuttlecock.

After going through the second expander, the steam is returned to a condenser that looks much like a standard automobile radiator. However, the Besler condenser is over twice as large as a regular one. In this case, the cooling air is gained mainly from outside air rammed over the large condenser metal surfaces rather than from a fan. Under conditions where the steam is too hot for cooling all of it to liquid, sensing units open a vent to release excess steam into the air, and this lost fluid is replaced by cool water from a reserve tank. Though the venting system only causes a small loss of fluid during normal operation, the reserve tank must be refilled at shorter intervals than other designs. On the other hand, a certain amount of weight saving and mechanical simplicity is gained in this approach.

Another difference in SE-124 details compared to the SE-101 is in the combustion and steam generator systems. These take the form of two large cylindrical containers. The steel tubing (common type steel and stainless) used in the generator is arranged in spiral pancake patterns stacked one on top of the other instead of vertical lengths connected by short curved joints at the end. The total length of the Besler spiral tubing is about 275 feet. The fuel and air are pumped into the combustion chamber in such a way that they mix while swirling around in a vortex pattern before the spark ignites them. The ignited gases flow over the steam generator tubes to make the steam, as occurs in the SE-101. The transmission used in the first test installation of the Besler was the conventional Chevelle three-speed type.

Though both the SE-101 and SE-124 now have been run for many hours at GM, the successful vapor-cycle car of the future—if a mass production design should ever evolve—may well be completely unlike either of them. An increasingly strong contender, for instance, is an engine that goes back to the rotating turbine concepts used in steam propulsion for electric generating plants. Lear Motors Corp., of Reno, Nevada, one of the newer organizations to work on advanced low pollution engines, has come up with several novel concepts based on the use of turbine wheels or helical screw units instead of pistons. Headed by famed industrialist-inventor Bill Lear, the firm began with a triangular-shaped, four-cylinder piston expander type design in the late 1960s called the Vapordyne. The firm installed one of these in a racing car and asked for permission to compete at the Indianapolis 500, so confident was Lear of the engine's high performance capability.

However, continued work showed the Vapordyne to be too expensive to build and, more important, too prone to failures of key parts. So Lear and his staff decided to look for other ways of going, including the use of multi-stage turbines and the substitution of a more efficient working fluid than water. After three years of intensive work, a

William P. Lear, head of Lear Motors, points to the 5½-inch diameter wheel used in the turbine that serves as the main power unit of the Lear vapor cycle engine.

radically new engine called the Learium Turbine System looked good enough for General Motors' president, Edward Cole, to visit the firm with his head of research laboratories, Dr. Paul Chenea. After looking at completed hardware and test results, they agreed to supply a Chevrolet Monte Carlo for a prototype engine installation as well as other assistance in the program.

As of the early 1970s, the Lear Motors design used a single shaft turbine, supplied by International Harvester, as its main power unit. The turbine wheel was 5½ inches in diameter and weighed only 23 pounds. Normal drive speeds for the wheel were in the 45,000-47,000 revolutions-per-minute range, but tests showed it could operate at speeds of over 90,000 rpm. without failure.

The use of a fluid other than water, called Learium, was

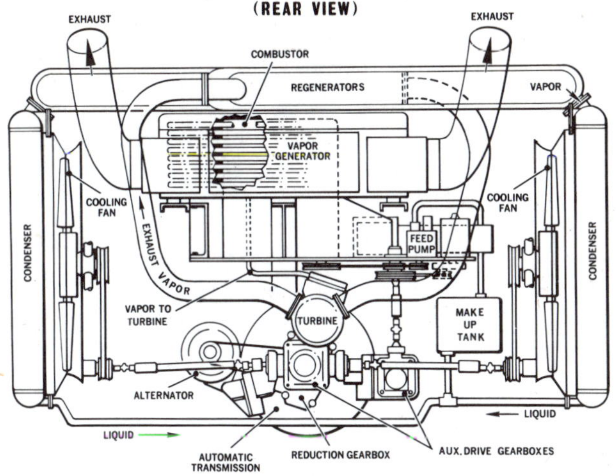

Diagram of the Lear steam engine installed in an experimental bus built for the San Francisco Municipal Railway system in a government-financed program. The Lear powered bus was one of three steam vehicles to go into operation on California routes in 1972. Goal of the program was development of a practical pollution-free bus power plant for use by the mid or late 1970s.

based on several things. For one, a major problem with all steam engine designs is the relatively low freezing point of water. Special care has to be taken in developing steam engines so that the danger of freezing is minimized by using special insulation and by making sure there is enough clearance in vital sections so freezing won't crack engine parts. By going to other fluids, the freezing problem can be overcome. However, most fluids examined to replace water have had other faults. For instance, a joint program between Thermo Electron Corporation, of Massachusetts, and Ford Motor Company involved the use of a fluid called thiophene. The performance of test engines looked good enough that the Department of Transportation awarded an experimental contract for continued work. However, a major stumbling block has been the ten

LEARIUM MAIN POWER TURBINE

Simplicity of the Learium engine main power turbine is indicated by the disassembled exhibit in the lower picture.

LEARIUM TURBINE

Cutaway through the turbine system of the Lear vapor turbine engine.

dency of thiophene to catch fire too easily if the engine temperature exceeds certain limits.

After studying many chemical compounds, Lear experimenters came up with a material based on the use of fluorine plus several other compounds that seemed to have less danger of fire. Because this organic fluid has a molecular weight several times that of water, it permitted use of a turbine expander much smaller than that needed for a steam design.

The engine intended for tests in the Chevy used a vapor generator consisting of small diameter tubing wound in a spiral around a central burner unit. The total size of this generator is about the diameter and thickness of a conventional passenger car tire. Starting is accomplished much as in other engines described in this chapter with the fuel-air mixture ignited to provide a temperature of about 700 degrees F. This is enough to vaporize the fluid, which is then led into an expander that looks like a helical screw. The helical grooves in this device have wider openings at the inlet end than at the outlet. Thus the vapor has to pass through smaller and smaller sections as it moves towards the turbine, insuring a great amount of compression of the Learium.

The compressed vapor hits the blades of the turbine with great impact, rotating the turbine which, in turn, transmits the rotating motion to the transmission—in this case a standard Chevrolet three-speed system. Some of the compressed fluid is diverted to drive a second, smaller turbine. This auxiliary turbine is used to run some of the accessory systems and thus lower the demands on the main turbine.

The engine uses a regenerator system that works on the same principle of heat transfer used in the regenerator wheels of the gas turbine of Chapter 4. In this case, the still-hot vapor leaving the turbine section passes through tubing that is located around the ducting from the condenser section to the vapor generator. Some of the heat from the exhaust vapor is transferred to the liquid going into the vapor generator. This action serves to preheat the latter while also cooling the vaporized Learium on its way to the condenser. In the condenser, cooling supplied by a large fan completes conversion of the working fluid from a vapor to a liquid for the next cycle.

Tests of the turbine in the early 1970s indicated it had low emission rates well under the proposed standards for clean air. The main turbine power output for the Chevy was a modest 110 hp., but the drive system could accelerate a 4,000-lb. car to 60 mph. in about 16 seconds. Bill Lear also stated the engine output could be upped to 250 hp. with only minor modifications to the main turbine and with no increase in size and weight. However, the tests indicated Learium caused problems of part corrosion that still needed to be solved.

In the early 1970s, then, things looked promising for vapor-cycle engines. However, it still remained to be seen if mass production engines could be developed low enough in cost and free of problems for widespread use. Also to be coped with was the weight problem; all the engines tested through 1971 were from 50 to 100% heavier than a conventional piston engine system.

6

The Electric Engines

The electric car probably has received the greatest publicity as the most likely answer to the pollution problem. After all, a battery-powered car doesn't burn any fuel, and none of its parts give off chemicals that would foul the air. In addition, there's no doubt that the principle works—there are already many special vehicles, from golf carts to small delivery trucks, that work on batteries.

Unfortunately, what seems like the simplest answer turns out to be far from simple when someone looks at the "big picture." It's one thing to have relatively small numbers of electric vehicles operating for short periods of time over short distances. It's something else to attempt to extend these systems to the operating needs of the vast majority of automobiles.

One of the problems with the conventional car battery is

that it can only provide electric power for a limited amount of time before it must be recharged. This is no problem in the internal combustion engine—almost all of the power to move the car is provided by the fuel explosions in the cylinders. In addition, the crankshaft rotation helps run the electric generator that regularly recharges the lead-acid battery while the car is running. If the lead-acid battery had to provide the primary power, many more battery units would have to be used and even then, the car might only go 10 or 20 miles before the battery system had to be recharged.

The lead-acid system, therefore, won't do for a practical, general-purpose passenger car. This rules out using a battery that is currently very cheap to buy. There are, however, many other possible chemical systems that can be used to build a battery. These other systems must be inexpensive, of course, but also must meet a number of requirements that are not demanded of the internal combustion engine battery. For one thing, they must have a high power density; that is, they must be able to provide a very high level of power (for short periods of time) for relatively little battery weight. This capability is needed for high energy demand periods, such as when the car goes to cruise speed from a stop position, or if fast acceleration is required to pass another car on the roadway. The second need is a high energy density. This means that the battery must pack a lot of energy into a lightweight package to provide the continuous energy supply that will allow a car to travel a reasonable distance—150 miles or more—without recharging.

It turns out that it's not easy to meet all of these requirements in one battery system. Some battery concepts have excellent power and energy density, but are much too costly for practical use. An example is silver-zinc, a battery

SEVERAL ELECTROCHEMICAL ENERGY CONVERTERS UNDER INVESTIGATION

Converter	Major Interest	Attractive Features	Problem Areas
Lead-Acid	Power characteristics	Excellent power density Compactness Economics and durability	Poor energy density Recharge time Poor power at part charge
Nickel-Zinc		Excellent power density Good energy density Promises good rechargeability	Limited cycle life Rechargeable zinc electrode
Nickel-Cadmium		Excellent power density Fair energy density Good cycle life Good rechargeability	High cost materials
Zinc-Air	Energy characteristics	Good power density Good energy density Low cost materials	Very limited cycle life Rechargeable zinc electrode
Organic Electrolyte		Excellent energy density Promising economics	Poor power density Durability unknown Rechargeability
Fuel Cell		Excellent energy density Rapid refueling Use of current fuels	Poor power density High cost materials Size and complexity
Sodium-Sulfur	Power and energy characteristics	Excellent power density Excellent energy density Promises excellent rechargeability	High-temperature operation Materials compatibility Durability unknown
Lithium-Chlorine		Excellent power density Good energy density Promises good rechargeability	High-temperature operation Materials compatibility Durability unknown

This table compares the properties of a number of possible battery systems being studied for electrically powered cars in the early 1970s.

that has been used in a number of spacecraft flights. The silver-zinc system can easily get cars up to 80 miles an hour and provide ranges of 80 miles or more on a full charge. But the amount of silver needed—not to speak of zinc— would push the cost for the battery alone to $25,000 per car. Besides this, it would take all the silver reserves presently available in the world to make enough batteries for the cars used in a single large city.

There are other battery ideas that don't have the cost drawback. But some have high power density without having good energy density, and vice versa. Others seem as though they may provide both of these properties without excessive cost, but may weigh more than all the rest of the car put together. Still others look good on paper, but a great deal more research and development must be done before engineers can be sure they can be produced in practical, mass production form.

The picture, then, is complicated, but it is not bleak and hopeless. The point is that much effort and years of study are needed until the all-electric car can once again compete on an equal level with some of the other systems described. We say once again because, just like the steam engine, in the early days of the automobile, battery-run cars were in the thick of the battle for the consumer market. Unfortunately, like steam, once the internal combustion engine took over the lead, the electric car research and development work almost completely stopped. It isn't easy to make up a half-century gap in research and development.

Before reviewing just how things stand with the electric car, it might be well to return to the question of pollution gains. The electric car doesn't pollute the air when it runs. However, for most proposed batteries, where regular recharging is necessary, there's another complication to be considered. Recharging is easy for the proposed electric cars—just uncoil an electric cord and plug into any available regular electric socket. But the power for these sockets, as for all things run by electricity, comes from large electric generating plants. These plants—as the brownouts in a city like New York have shown—already are straining to meet demand. New ones are being built to take care of expected increases in home electric volume. If millions of cars were added to the load, still more plants would be needed.

But electrical generating plants also are being criticized for contributing to pollution. Many of these plants burn coal and other fuels that can cause undesirable chemicals to go into the air from their smokestacks. Going to cars with rechargeable batteries might just transfer the pollution problem from the car's tailpipe to the generating plant's smokestack. Curing pollution requires considering a great many sources of dangerous substances. Perfection of a practical battery-run car would increase the need for developing non-polluting electric generating plants.

There are some other ways with battery-run vehicles to cut down on the need for using outside electric systems. One of these is to use a device called a fuel cell. A fuel cell works along the same general principles as a battery. However, instead of generating electricity with materials that can't be replaced once they have been used up, the fuel cell works with fluids that are replaceable. In a fuel cell, once the electric generating capacity of certain materials has been used up, the cell can be brought back to full power just by adding new amounts of these materials. The fuel cell is refilled in the same way that a gas tank is, but the chemical reaction in the cell supplies electric power directly to run the car, whereas gasoline must be exploded in a car's cylinders to do this.

Another approach is to use a battery device as part of a

combination system. One such system is the Stir-Lec car, which will be described in the following chapter. However, there are many other combinations being studied, including combinations of batteries with gasoline engines, or multiple battery systems using one or more advanced, non-reusable batteries matched with a fuel cell. In the hybrid systems, the idea is to take advantage of the best properties of both power sources while minimizing each one's disadvantages. For example, the gasoline engine or fuel cell might provide the power for general cruise operation at a relatively constant speed, with the battery cutting in when sudden peak loads are needed.

The system is designed to recharge the batteries while the other engine runs the car—much as the batteries in conventional internal combustion engine cars are recharged while the car is moving. With this approach, the need to tax the outside electric generating system by plugging the battery into a wall socket is eliminated or at least greatly reduced. The main difference between a hybrid battery system and a regular car of today is that the battery is not just a supplementary power source to help in starting the car or to run the radio or other extra equipment at times. It is also used to give primary power to run the car for certain periods of time.

At this point, we might take a look at the kinds of batteries being considered as eventual substitutes for the old lead-acid standby. (The fuel cell is a special case and will be discussed a little later in the chapter.) Nickel and zinc are among the materials that have been examined in several forms. Batteries using them include: nickel-zinc; nickel-cadmium; and zinc-air. All have advantages and disadvantages. In nickel-cadmium, for instance, as in silver-zinc, materials costs are very high, but at this writing, the zinc-air system seems to have the biggest margin of possible advantages over possible disadvantages and has therefore been widely studied. Another class of batteries that shows promise makes use of a molten salt electrolyte in combination with highly energetic reactive chemicals. Possible molten salt batteries include sodium-sulfur and lithium-chlorine.

The reference to an electrolyte brings up the subject of how a battery works. The main parts of any battery include two electrodes either of different materials or with different electrical "charges," and a substance into which these extend, called electrolyte. The electrode materials are chosen so that one will be positively charged during the cell's operation and the other negatively charged. (A single battery section is called a cell. For increased electrical output, a battery may contain two or more cells interconnected inside the battery housing.) One electrode, called the anode, is used to introduce a flow of electricity into the battery and the other, called the cathode, is the one from which electrically charged particles leave the cell.

The electrodes are not physically connected to each other. The common tie between them is the electrolyte. This is a chemical compound that conducts electricity when it is dissolved or melted. The electrolyte can take various forms. It might be a liquid solution, such as in the conventional lead-acid car battery. Or it might be a solid—or semi-solid—as in a flashlight dry cell. In the simplest systems, such as a Voltaic cell with a zinc cathode and carbon anode—or the lead-acid battery using lead plates that are alternately positively and negatively charged—the electrolyte is a weak solution of sulfuric acid in water.

The sulfuric acid, when dissolved in water, separates into atoms with both postive and negative charges. (A positively charged particle, or ion, has less electrons than it normally has, while a negatively charged particle has more electrons than normal.) When an outside current is introduced into the battery by turning a switch, such as a car ignition key, particles interact with the electrodes to give up some electrons in one case or take on electrons in the other. The chemical reactions involved are beyond this book's scope, but the net result is that once the outside current starts this process, the reactions in the battery continue, either until one of the electrodes is eaten away, or the switch is turned off.

In zinc-carbon and the lead-acid systems, the chemical reaction takes off some of the zinc electrode or lead plates each time the battery is used, eventually destroying the cell. The process can be reversed to a great extent by recharging—that is, reversing the polarity of the electrodes so that the electrodes return to almost their original chemical state. Recharging is one reason why lead-acid batteries can last for several years of normal automobile operation. However, the process never completely restores all the chemical substances to their unused state. A time does arrive when further recharging won't restore a battery of this kind to good condition.

The situation is different, though, for a fuel cell. In this case, the electrodes act to introduce and extract current. from the battery. But they do not take part in the chemical reactions needed to create a flow of electricity in the cell. All of the necessary chemical reactions occur in the fluids poured into the fuel cell. As a result, a well-designed fuel cell theoretically can have an almost unlimited life.

In the zinc-air cell, the anode is a zinc plate; the cathode is a source of oxygen (an air cathode is a porous structure designed to allow air to enter the cell but prevent the electrolyte from seeping out); and the electrolyte, in the systems that have been most studied, is potassium hydroxide. The air electrode is made of carbon, a plastic material such

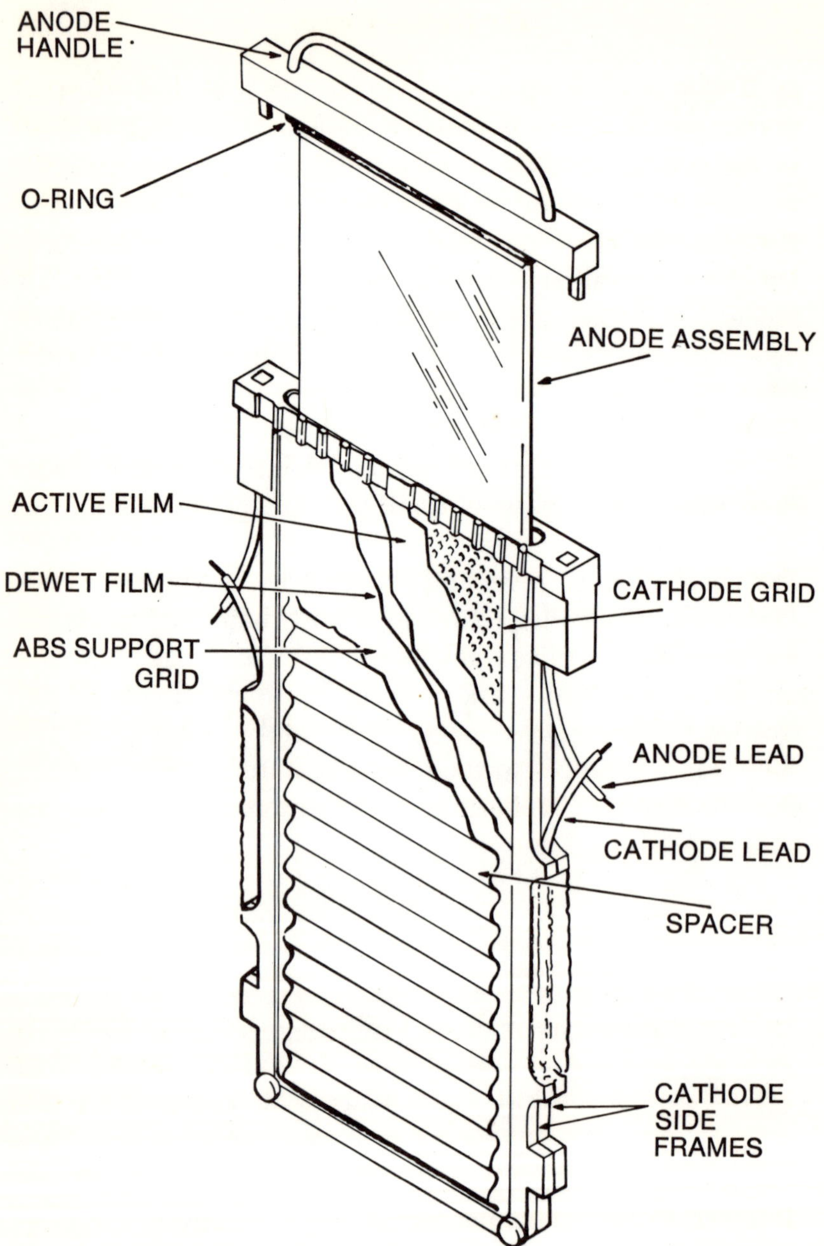

This engineering cross-section diagram shows the main parts of a typical cell assembly for a mechanically rechargeable zinc-air battery.

as Teflon and a special chemical catalyst placed on a porous metal matrix such as nickel. Electricity is generated in the cell by combining zinc and hydroxyl ions from the electrolyte to form the compound, zinc hydroxide. This reaction releases electrons that flow from the output electrode to provide the desired current to run the car's systems. The hydroxyl ions (charged particles consisting of one atom of oxygen combined with one of hydrogen) are released by a reaction between incoming electrons from the circuit, and water and oxygen provided by the air.

The design of a zinc-air cell is more complicated than a lead-acid system or a silver-zinc battery. But there are several reasons why many research groups are working on zinc-air principles. James V. Leonard, of McDonnell Douglas Corp., St. Louis, Mo., and Professor Herbert A. Crosby, University of Missouri, put it this way:

"There are many types of batteries available on the market today, so why select one that is still in the research and development stage? As an energy source for the automobile, major factors that must be considered are:

> Specific energy (watt-hours/pound)
> Specific capacity (watt-hours/cubic inch)
> Economics
> Air pollution.

"Our studies, as well as those of other experimenters, indicate zinc air potentially has advantages in all these areas (except the last, since all batteries, as prime power sources, are not in themselves pollution problems)."

Though too expensive in the early 1970s, it had the promise of eventual low cost since it can be made from inexpensive materials. In specific energy (which is the same thing as saying power density), it could provide 100 watt-hours per pound for a typical system; while equiva-

Lightweight zinc-air battery is shown installed in an experimental gold cart built by GM Laboratories.

lent silver-zinc cells would only give about 75, lead-acid under 25 and nickel-cadmium about 15. Similarly, in specific capacity (that is, energy density), zinc-air has an advantage edge over its main battery competitors. Where a typical lead-acid cell would have a capacity of 3 watt-hours per cubic inch, its zinc-air equivalent would have 7 watt-hours/cubic inch. Silver-zinc is better than lead-acid (about 5.5 watt-hours/cubic inch) but still well below zinc-air.

Three kinds of zinc-air batteries are being developed: disposable types, mechanically rechargeable and electrically rechargeable. As examples of the first class, Yardney Electric division of ESB, Inc., has been testing a 180-ampere-hour device using a very low cost carbon material as the air cathode. American Cyanamid Company has researched a system in which zinc powders are bound together by a gelatin-like material containing the potassium hydroxide electrolyte. These batteries weigh about 200 pounds and would have to be replaced every week or two. (However, the normal gasoline-fueled car takes on over 200 pounds of fuel every week under normal use conditions.) The goal is to develop disposable zinc-air cells that wouldn't cost much more than a week's supply of gasoline. The disposable cells wouldn't be destroyed, but would be traded in on a new cell. The depleted cell would be sent back to the factory for reworking with fresh materials and resold.

Mechanically rechargeable cells are designed with a replaceable zinc anode assembly. Some batteries of this kind are already on the market for industrial uses. Eagle Picher Co., for instance, produces a cell with a maximum capacity of 100 watt-hours/pound, output of 150 ampere-hours and life of 125 cycles. Energy ratings for mechanically rechargeable types, at the moment, seem to be more in line with the power supply needed for electric cars than disposable types. However, still greater specific energy and capacity are required for practical, general purpose cars of the future. In the meantime, the types produced thus far are much too expensive for economic automobile applications.

The electrically rechargeable battery, Mr. Leonard and Professor Crosby believe, has the best potential for both low cost and easy maintenance for tomorrow's electric car. A number of companies here and abroad are testing these systems in experimental cars. Sony Corporation of Japan has been road-testing a car with a zinc-air cell using a circulating electrolyte containing powdered zinc. The electrolyte can be collected and chemically reprocessed so the materials can be used over again. American firms doing this kind of work include McDonnell Douglas, Gulf General Atomic (San Diego, Cal.) and Leesona-Moos. Most of these tests have attained battery recycle totals of from 20-40 up to over 100. However, for a zinc-air battery to be competitive with a gasoline-powered car, it would have to be capable of 1,000 cycles, the number needed for a lifetime of 50,000 miles. So, for all its promise, the zinc-air system still has a long way to go.

Molten salt batteries promise several advantages over zinc-air types. These include greater power and energy density and the likelihood of excellent rechargeability. The major drawbacks include the need to develop inexpensive material systems that can work at temperatures well over 1,000 degrees F. higher than zinc-air or lead-acid cells. Still to be answered, as well, is how long and how reliable a molten salt system can work without failure.

General Motors has concentrated a great deal of research effort on a molten salt system using lithium and chlorine. This battery uses a cylindrical graphite chlorine electrode surrounded by a lithium electrode. The space between the cells is filled with fused lithium-chloride which acts as the electrolyte. When the cell is ready to use, its temperature rises until liquid lithium floats on molten lithium-chloride and saturates the wick leading to the lithium electrode. Chlorine gas needed to complete the reaction is supplied to the chlorine electrode from a storage tank.

A complete lithium-chlorine cell has been tested for many hours in GM laboratories. The cell has gone through several changes until a design is now in operation with a lifetime of over 3,000 hours. This is excellent progress, but to operate, the cells made so far require heat input from special laboratory equipment. The next step requires designing a molten salt battery that can generate the needed heat energy in a lightweight package suitable for installation in a vehicle.

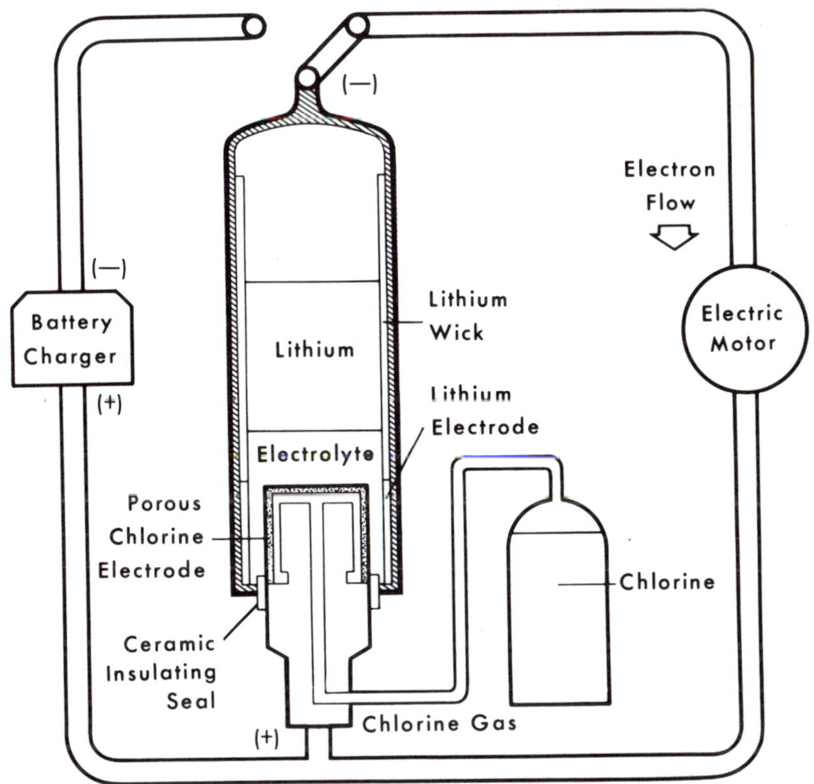

Operating diagram of the General Motors lithium chlorine cell, showing the battery in the discharging mode.

Fuel cells, as has been pointed out, differ from conventional batteries in that they can operate continuously as long as fuel and air are available, and can use hydrocarbons (or some derivative of hydrocarbons, such as hydrogen) for fuel. Fuel cells also have received a great deal of attention because of their adaptability for space missions. The National Aeronautics and Space Administration (NASA) has given contracts to a number of companies to build fuel cells for manned flights. Operating power

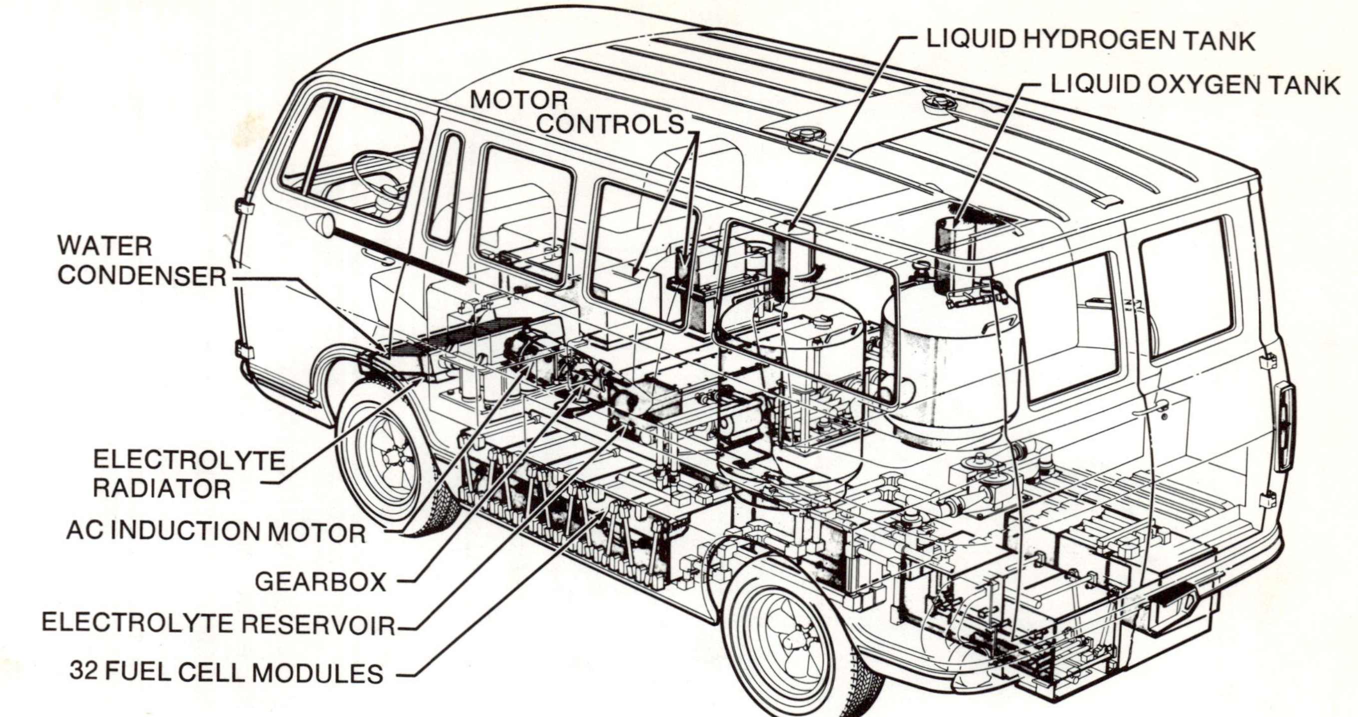

Cutaway view of the fuel-cell-powered Electrovan shows that there is little room left in the van after the experimental fuel cell system is installed.

sources of this kind were used as vital parts of all the Apollo moon flights.

However, though fuel cells have thus been operational for quite some years, they are much too expensive in their current form for use in ground vehicles. They also take up too much space in a conventional car or truck to leave room for passengers or cargo.

As an example, a fuel cell power system derived from Apollo technology has been installed in GM's Electrovan, a vehicle roughly the same size as a Volkswagen bus. The total weight of the Electrovan is 7,100 lbs. but more than half of this weight (3,930 lbs.) is taken up by the fuel cell power plant and electric drive system. This much equipment is needed to give the vehicle a top speed of 70 mph., acceleration from 0-60 mph. in 30 seconds and a range with full tanks, of about 150 miles.

The power plants use 32 thin-electrode fuel cells connected in series. The cells, built by Union Carbide Corp., supply a continuous output of 32 kw. and peak output of 160 kw. The materials needed to operate the cells are liquid hydrogen and liquid oxygen. These liquids are stored in special pressurized tanks that maintain inside temperatures hundreds of degrees below zero. These temperatures are needed to keep the oxygen and hydrogen cold enough to be in liquid form rather than gaseous. The hydrogen-oxygen cell produces current in reverse of the way that a conventional battery cell does. In a battery cell, the process of electrolysis is used, in which water is separated into separate components by passing a current through it. In the hydrox system, the hydrogen and oxygen flow into the cell and react together to form water, liberating energy in the form of electricity in the process.

Fuel cells can be built with starting materials other than liquid oxygen and hydrogen. General Electric, Yardney

Lineup of four experimental urban cars being tested at GM's Technical Center. Car at the far left is a three-wheeled commuter car, the next car is a 512 hybrid gasoline-electric design, the third a 512 electric and the last one a 512 gasoline-powered design.

and dozens of other firms are experimenting with all kinds of fuel cell substances in hopes of coming up with a more compact, lower cost battery for mass production application.

Before any of these individual battery devices become the sole power sources for cars, hybrids are liable to fill the gap.

All car manufacturers, plus electric equipment firms such as Westinghouse and General Electric, have been building prototype cars with hybrid engines. The idea is to combine the uses of the engines for reasonable performance but with much lower output of pollutants. The GM XP-883, for instance, can work in either all-electric, all-gasoline or hybrid modes. This special purpose commuter car is accelerated from 0-10 mph. by the electric

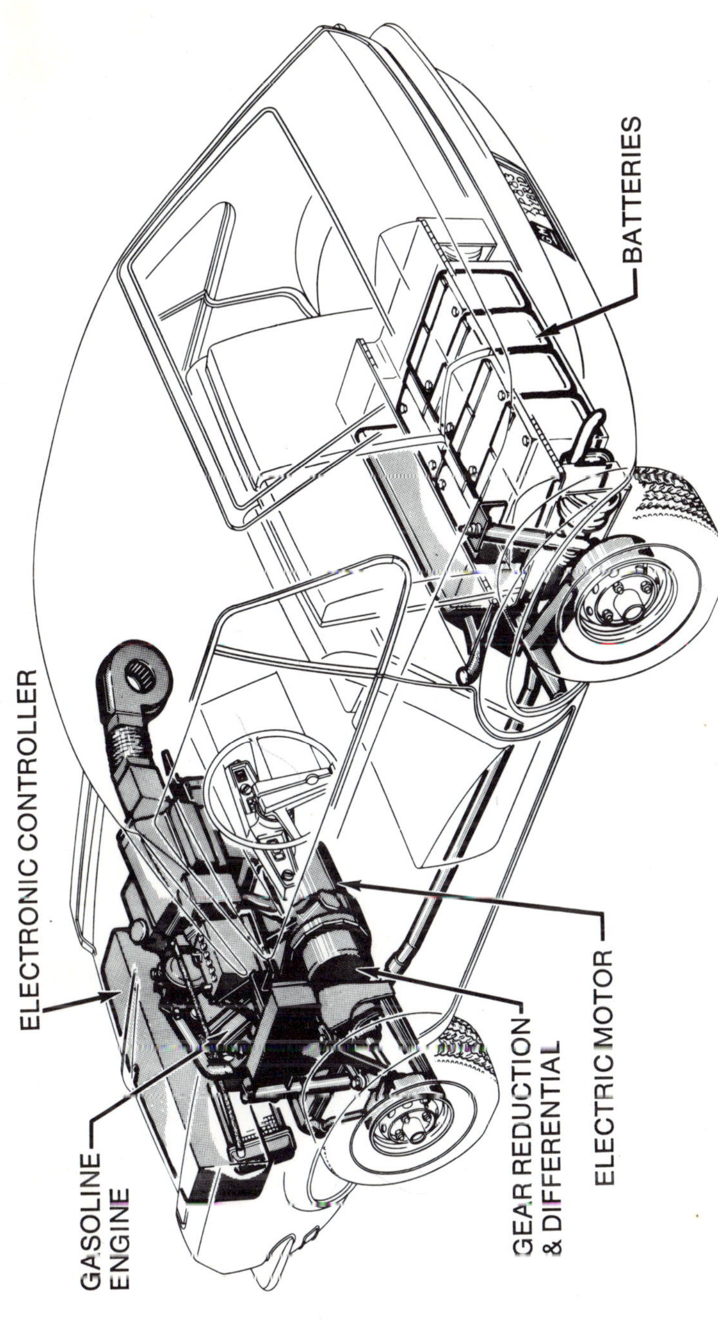

Installation diagram for the hybrid gasoline-electric engine system in the XP-883 test car.

motor (powered by six 12-volt batteries providing 72 volts in all), after which the gasoline engine starts and provides power for steady speed operation. The gasoline engine recharges the batteries when it's running as in any conventional car. If the driver has to accelerate rapidly, the two engines work together to provide much higher peak power output.

7

Stirling, the Quiet Engine

The car of tomorrow may not be powered by any of the advanced engine concepts that hold the spotlight today—such as the gas turbine, the Wankel or the electric motor. It may come from engines in very early stages of study, such as the free piston or the magnetogasdynamic systems described in the next chapter. Or the eventual winner might be engines that are already being tested in full-size automobiles, such as the Stirling. In fact, if noise pollution regulations should ever become as stringent as those covering chemical pollutants, the Stirling or the vapor-cycle engines described in Chapter 5 might come into their own, because these are the quietest types of designs now being studied.

Of course, the Stirling, like the vapor-cycle idea, has disadvantages as well as advantages. The basic approach of an

111

engineer to any design has to be one of compromise. H. G. Warner, Executive Vice President of General Motors for Engineering, pointed this out to newsmen a few years ago at a meeting reviewing alternate power plants.

"Engineering," he said, "combines scientific principles with economic reality. One might think that engineers could design a perfect mechanism if they gave it enough study and research, but they can't. People's needs and wants change. Materials change. Processes change.

"This is quite apparent in a classic advertisement that appeared back in 1912. In a bylined ad, R. E. Olds (for whom the Oldsmobile is named) said of his Reo automobile: 'I do not believe that a car materially better will ever be built. I test my gears with a crushing machine—not a hammer. I know to exactness what each gear will stand. I put the magneto to a radical test. The carburetor is doubly heated, for low grade gasoline. So in every part. The best that any man knows for every part has been adopted here. The margin of safety is always extreme. I regard it impossible at any price, to build a car any better.' His price was $1,055—without top and windshield, without self-starter . . . and of course, without automatic transmission, radio, four-wheel brakes, air conditioning, tilt wheel, electric windows and many other features. But still he felt it was impossible to build a car any better."

Warner noted that all engineers are aware today, if they weren't always in the past, "that everything and anything can be improved." He always stressed that constant review of all ideas is necessary for several reasons. For one thing, "many of the big breakthroughs were not recognized as such. One of these was the pneumatic tire. It wasn't even developed for the automobile. An Englishman, Robert Thomson, patented it in 1845 (when there were no automobiles)."

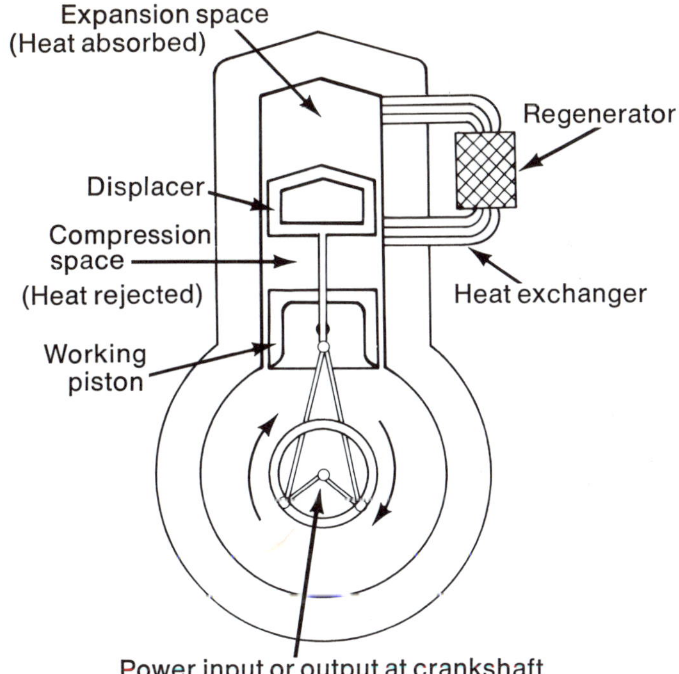

Basic parts of a Stirling cycle engine. Heart of the design is the regenerator, a device in which hot gases leave some of their heat, which is later picked up by cold gas passing through the unit. In one working cycle, the working piston first moves upward, compressing gas in the compression space. The displacer then moves down, forcing gas through the regenerator in the expansion space. While the displacer is down, the hot gases expand to push the working piston down, causing rotation of the output shaft.

A second reason, Mr. Warner pointed out, is that ideas must be considered in the context of their time. They may be born too soon . . . rejected . . . and then resurrected as advanced engineering, materials, plastics, manufacturing techniques, mass production and market acceptance make them possible."

If the Stirling should ever become widely used in cars, this last statement would surely apply. The principle was proposed over 150 years ago by a Scottish minister, the

Reverend Robert Stirling. His idea, presented in 1816, was to use air as the expanding substance of a heat engine. The concept had the advantage of using a readily available substance—air—to move the working piston and also was a fairly simple system.

Stirling's original system consisted of a brick furnace fueled with coal or other conventional fuel, a metallic shell supplied with air from an outside source, a movable displacer, and a tube connecting the displacer to the working cylinder. The furnace was designed to supply enough heat to the shell to heat the air to a high temperature in an enclosed area. After a certain period of time, the hot air was allowed to expand into a chamber and exert pressure against the bottom of the displacer. The movement of the displacer forced a high pressure stream of air through the tube and against the face of a piston. This forced the piston and its working shaft (on the other side of the piston) to move up—or out—into the cylinder. After the heated air had given up its energy to the displacer, it passed into the atmosphere through exhaust pipes. With the removal of the expansion pressure from the underside of the displacer, the spring-loaded piston moved back down the cylinder, causing the displacer to return to its original position until the next surge of heated air started the cycle again.

It can be seen that, like the steam engine, this is an external combustion engine. That is, the combustion step takes place outside the working cylinder. This means that it is possible to design such power plants as closed-cycle types in which the working fluid is continuously reused. This was not done in Stirling's original engine, but was adapted in later systems. It is a major advantage of the engines that have been examined by experimenters in recent years. In addition, any fuel that will supply the needed heat can be used, combustion is continuous, and thus almost no pollution results from a properly designed Stirling engine.

After presenting his original ideas in 1816, Stirling continued to work on improved versions, including types that used portable burners rather than furnaces for heating. His most important contribution in his later work occurred in 1827, when he demonstrated the use of heat regeneration. By using thin strips of metal he took some of the heat from one section of the engine, providing a cooling effect on the air in this region. He transferred the heat to the combustion section, increasing the heat energy of the air in this area. The principle of regeneration has been important in many engine systems, as was noted in earlier chapters on gas turbine and vapor-cycle engines. In the intervening years from 1827 to recent times, Stirling's idea of heat transfer has been important in many applications having nothing to do with engines. These include, for example, furnace systems used to melt and refine metals or to prepare metal alloys.

Other inventors were excited by Stirling's work and attempted to develop practical designs for various applications. One of these was Swedish-born engineer John Ericsson, who built a large hot air engine (he called it a "caloric" engine) to power a 2,200-ton ship. The system worked, but never competed successfully with other ocean-going vessels. Ericsson had long since abandoned this approach by the time he designed the famous ironclad warship *Monitor* for the U.S. Navy in the American Civil War.

Stirling's engine won particular attention for a few decades after its invention because it was much safer to operate than the steam engine, which was then also in its infancy. After a while, when experimenters learned how to properly design steam parts, such as high pressure boilers that could take a lot of stress without exploding, and parts that wouldn't leak scalding steam, new applications overwhelmingly went to steam rather than to the air engine.

One of the main problems was the low density of air compared to other working fluids. Because there are much fewer air molecules in a given volume than water or other fluids, the air engine ended up being much bulkier than competing systems. An even more serious drawback is the fact that air is not as easy to heat to a high temperature as is water. Therefore, there is a much greater loss of heat to the exhaust system than there is in a steam engine. Another drawback was the difficulty in finding materials for the shell that contained the air being heated, because of the extremely high temperatures involved. To achieve the amount of heat transfer for the burner through the shell into the working air, the flame temperature had to be higher than that needed even in a high pressure steam system. The temperature had to be so high, in fact, that while some early engines didn't explode, they sometimes stopped working because large sections of the metallic shell just melted away.

Though a few experimenters continued to build small Stirling-type engines off and on in the years after Stirling's death, the engine was all but forgotten for close to 100 years. In the meantime, though, major advances occurred in high temperature materials, cooling systems for such materials and in the understanding of regenerating equipment. Some engineers and scientists then realized that these advances might make a new version of the Stirling system practical.

The first attempts on any sizable scale began in the Dutch firm of N. V. Philips Gloeilampenfabrieken in the late 1930s. Work was interrupted by World War II but was resumed again in the 1940s. Research data from these studies indicated a modern Stirling engine could be built with high efficiencies. For instance, working models of some engines were built with thermal efficiencies of close to 40%, better than diesels and twice as good as conventional gasoline internal combustion engines.

In the post-World War II years, many other researchers took another look at the Stirling concept, including laboratories of the major U.S. car manufacturers and such independent research organizations as Battelle Memorial Institute, Columbus, Ohio. The Stirling concept would work with gases other than air, promising a system more compact and easier to raise to working temperatures, and many methods of heating might be used, from nuclear energy to ordinary sunlight.

Battelle's early work was based on the use of sunlight as a heat source. The concept was aimed at providing, not a car power plant, but something that might be used in remote regions of the world where heat sources aren't easily obtained. However, it also served as a start towards gaining new understanding of the processes involved in improved Stirling designs.

In this system, the displacer, called a regenerator piston, served a dual role. It accomplished both compression of the working gas and the heat transfer function needed for maximum efficiency. The regenerator piston was installed in a cylinder having a transparent dome at the top. This dome was filled with highly heat-absorbent steel mesh blackened to increase this property. The piston itself was hollowed out and filled with steel wool. The dome-shaped top of the piston and its walls were solid metal, but the bottom metal plate was pierced by a great many tiny holes. This allowed some gas to move up into the inside of the piston to either pick up heat transferred through the dome or give up heat to the steel wool.

As in Stirling's original engine, the displaced chamber was connected by a tube to a second chamber containing a working piston. This was a closed system with seals used to insure that a certain amount of gas was constantly maintained in the space between the two pistons. The working piston was connected to a crankshaft, one end of which also was connected to the underside of the displacer by a long rod.

In operation, the first step was absorption of heat from the sun by the steel mesh in the dome. This energy was transmitted through the rounded top of the displacer to the gas inside this hollow piston. As the gas heated up, it expanded. The displacer was driven downward and compressed the rest of the gas sealed between the displacer and the top of the working piston. The pressure on the working piston, as in Stirling's original engine, forced it down, causing a crankshaft to rotate.

Unlike early engines, the downward movement of the crankshaft caused an upward movement of the displacer until its top was back against the steel mesh of the dome. The inside of the displacer, during this operation, performed a very efficient heat transfer step. During the expansion of the gas in the previous cycle, some of the heat was given up to the steel wool inside the displacer. When the cool gas moved back into the displacer as that part moved upward, it picked up the heat energy that had been left in the steel wool. As a result, the gas didn't have to pick up as much new heat from the sun to start its next expansion stroke. The regenerator action of the steel wool insured a smoother and a more efficient running of the engine.

The Battelle program used air as the working gas for some of its tests. However, in many runs quite a few other gases were substituted, including hydrogen and helium. These gases can take up heat energy in a more practical fashion than air. Most Stirling work at other research institutes, and in industry in recent years, has concentrated on using such alternate materials. Promising engine systems for passenger cars and Army ground vehicles have been built and tested with Stirling engines using nitrogen in one case and hydrogen or helium in several others.

The use of gases other than air as working fluids has been one step forward in Stirling engine performance. Another major goal of experimenters has been to develop new and better sources of heat energy. In most cases, carefully engineered burners using diesel fuel have served this purpose. But many other energy storage methods are being considered, including "heat batteries" and heat supply systems based on the energy released from certain kinds of chemical reactions. Some scientists have looked into systems that might use a nuclear source, such as a small container of heat-emitting radioisotopes to expand the working gas, but this seems unlikely for car applications for the immediate future.

An example of a chemical reaction system studied at General Motors and other Stirling engine research groups is one using the metal, lithium, and a material called Freon 114. In this case, the reaction between the two substances results in a stored energy in watt hours/pound of 1,345 compared to only 10 for a normal lead-acid battery used in the production cars of the early 1970s. The conversion efficiency of the chemical reaction is only about 40% (that is, the system has an energy output of 540 watt hours/pound) for a Stirling engine. By comparison, the lead-acid battery has an output of 9 of its 10 watt hours/pound of stored energy, or 90%. On the other hand, the Stirling reaction system gives many times the energy output needed to run an engine than does lead-acid. Heat batteries may have much lower stored energy levels, but they still can provide

Heated aluminum oxide was used as the main material in the "heat battery" used for the Stirling engine that operated this experimental "Calvair" car.

outputs from 10 or 15% up to 500-1,000% that of regular batteries.

In 1964, General Motors built and operated a heat-battery Stirling engine in a modified Corvair passenger car. The system is based on the use of a molten salt which slowly cools as it feeds heat to the engine. The heat-battery material used here was aluminum oxide, or alumina. This is the same material used to make ceramic pottery and other kitchenware. In the system, natural gas was heated in a special burner to bring a tank of aluminum oxide pellets to very high temperature. The burner turned off automatically when a detector indicated the pellets were hot enough, and turned on again when the pellets had cooled to a temperature lower than desired for engine operation.

Heat from the pellets was transferred to nitrogen gas circulating around the alumina tank. The nitrogen then carried this heat to the engine, where it raised the temperature of the working fluid to about 1,200 degrees F. This fluid—hydrogen gas—then expanded to run the engine as in previously described systems. The car powered by this system, called the Calvair, performed well on many test runs over the roads at the GM Technical Center in Detroit. It demonstrated the principle, but the engine was too heavy (roughly twice the weight of a gasoline engine of equal power) and too expensive for a practical production car application.

However, the fact that the car ran very quietly and with relatively low pollutant emissions was another boost to Stirling work at GM and abroad. The data from the Calvair encouraged new studies of energy storage systems. One of these was a compound called lithium-fluoride which promised temperatures from 800 to 1,700 degrees F. (slightly below the 800-2,100 degrees F. range of alumina), and chemical reaction systems involving alkali metals and halogen gases. (The alkali metal group includes such metals as sodium, potassium and phosphorus. Examples of halogen gases are argon, neon and helium.)

Despite the performance possibilities of some of the more unusual fuels, the best compromise between engine cost and good efficiency would be to use more common heat sources, such as kerosene or diesel fuel. The main consideration is whether systems using such heat sources can meet air pollution standards. Preliminary tests of engines built both in the U.S. and at Philips Research Laboratory, The Netherlands, seem to give an optimistic outlook to date. A joint licensing agreement for Stirling equipment between General Motors and Philips was signed in 1958.

Philips, for instance, built a single-cylinder Stirling engine with a rating of 80 horsepower by the late 1960s that has been operating with a variety of different burners. Typical burners are the slit type and the louvered design. Both burners consist of large, cylindrical metal housings that look like 10-gallon cans with a large inlet at the top for injection of diesel fuel. The slit type has a thick metal wall pierced by long, vertical openings through which air is taken in for combustion. The louvered system uses a series of smaller slits arranged along sloping lines around the burner periphery.

Operating tests showed that both of these burners provided almost complete combustion of the fuel. Emissions of unburned hydrocarbons and carbon monoxide were well below the proposed U.S. maximum permissible values for the years after 1976. However, the tests also underlined the importance of burner design, because the emission levels of the slit-type chamber were considerably below those for the louvered design.

General Motors compared data for the Philips engines with its own design, a smaller engine (single-cylinder) rated at 10 hp. at 3,000 rpm. The GM engine was operated with No. 2 diesel fuel and with the hydrogen working fluid sealed inside the engine. The burner for the GM engine looked quite different from the Philips types. Made in two parts, it looked a little like an old-time metal helmet worn by footsoldiers of the Middle Ages. The domed top section was fastened to the hat-shaped lower part by six round, hollow bushings. (The openings in the bushings served as air entry holes.) In addition, long slits in the sloping dome allowed cooling air to enter the chamber in a swirling motion that improved fuel mixing and insured better burning. An atomizing fuel nozzle, with an igniter tip projecting from its bottom surface, fitted into a hole in the top of the dome to complete the combustion system.

Tests of the GM system (the GM engine section used a two-compartment cylinder with displacer in one compartment and working piston in the other, as in most Stirling types discussed so far) showed very quiet performance and low hydrocarbon and CO emissions. However, the GM engine seemed to have slightly higher emission levels than the Philips slit type at an air-fuel ratio of 20-1.

All of the engines, when run without any special controls or modifications, tended to have oxides of nitrogen outputs as unfavorable as in spark ignition engines. The main reason was that the highly efficient burning process in the Stirling systems used up practically all the fuel, but in so doing caused high combustion temperatures. As we have mentioned earlier, high temperatures are the factors that start the chemical reactions that produce nitrogen oxides.

Several methods of solving the problem were tried. One was to reduce the amount of air preheating used in the burner operation. In runs with the air preheater removed completely, for instance, a fivefold decrease in nitrogen oxides resulted. A second approach was to use higher air to fuel ratios (that is, the mixture was thinned) and retain full preheat. A third approach was to recirculate part of the cooled exhaust gas. This reduced oxides of nitrogen in three ways: the recycled gas lowered the flame temperature; the recycling increased the flow rate through the engine so that the fuel-air mixture was exposed to the hot chamber temperatures for a shorter time; and third, the exhaust gas had less oxygen to react with nitrogen. The disadvantage of any of these ways of lowering oxides of nitrogen output was to reduce the overall thermal efficiency of the engine. However, all of the work indicated that engines with acceptable performance and pollution emission properties could be developed.

By the early 1970s, Stirling engines of many sizes and powers had been successfully demonstrated. In 1971, Philips had installed a 200-horsepower, four-cylinder engine in a standard bus. To date, this engine system, installed under the bus floor, has been used for thousands of miles over regular bus routes with excellent operating results. The fuel used is regular fuel oil, such as that used for home heating with helium as the working fluid. Starting time, from switching the ignition key to "on" position to getting the cylinder temperatures to the desired 700 degrees C., is 20 seconds. Acceleration under no-load conditions, from idle to the engine's maximum rotating speed of 3,000 revolutions per minute, is only 1/10 of a second.

In addition to this work, Philips, GM and other firms built other engines able to reach horsepowers as high as 400. Philips also began experiments in the early 1970s on an engine system using a series of interconnected cylinders in such a way that no displacer was needed. In this system, heated and cooled volumes of gas were interchanged from one cylinder to the next to move the pistons up or down.

The variety of different Stirling systems under study at

this time promises many applications other than for ground transportation systems. Among the alternate uses proposed are long duration systems for underwater vehicles. These deep-diving craft someday might help bring resources from the sea floor to replenish dwindling land supplies of vital metals and minerals.

Despite the work on individual Stirling engines, though, their first practical applications in automobiles might be as part of a hybrid engine system for use in cities. Because the most vital immediate need in the fight against pollution is in the concentrated city areas, engineers have been looking at special cars with limited top speeds, range, etc., and very low noise and chemical pollution outputs. Several programs were begun in the late 1960s and continued into the 1970s, combining a Stirling engine with an electric-battery power supply. This kind of hybrid, engineers pointed out, promised the advantage of using a heat engine of small power rating to run the car at steady speeds, with peak power demands provided by a supplementary electric drive system.

GM has run several experimental cars in the "Stir-Lec" series. These are modified Opel Kadett cars with a Stirling engine (using hydrogen gas) in the trunk area and lead-acid batteries under the front hood with various electric systems (alternator and induction motor plus electric controls) connecting the two-power sources. In normal operation, the Stirling engine causes high pressure hydrogen to rotate a turbine wheel that operates an electric alternator. The electric output from the alternator supplies power to an electric drive train that rotates the car's wheels.

If the power demands suddenly rise, electronic circuits sense this and turn off the Stirling engine, transferring the electric circuit network so that it draws power from the batteries. As soon as the driver reduces speed, the same

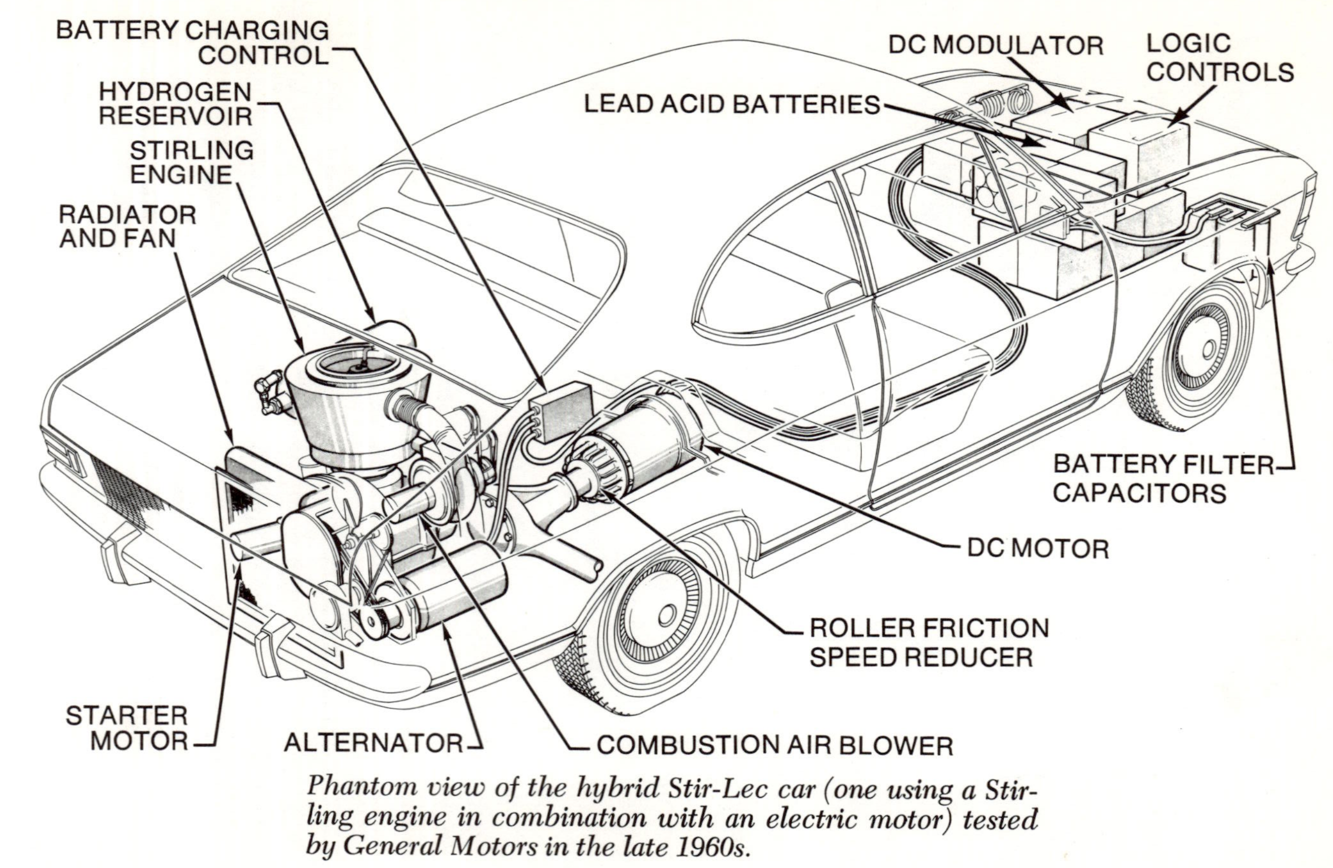

Phantom view of the hybrid Stir-Lec car (one using a Stirling engine in combination with an electric motor) tested by General Motors in the late 1960s.

system cuts the batteries out and switches the Stirling system back into operation.

It's a long way from the experimental work described here to a production engine, either Stirling alone or a hybrid. However, the low pollution properties of this engine may well mean that, sooner or later, "there's a Stirling in your future."

8

"Way Out" Power Plants

The engine that powers the car of the year 2000 may look like one of the systems already described in this book. Certainly this is possible, since almost all the experimental engines reviewed in earlier chapters have advanced to the point where practical production designs could evolve in ten or twenty years. But the car of the 21st Century might be powered by any number of different engines based on approaches that have either received only limited attention or have mainly been studied in small test setups in the laboratory. These "way out" ideas range from engines using relatively conventional pieces of hardware, such as the free-piston concept, to highly novel possibilities, such as electrogasdynamics or nuclear power.

Actually, for a while in the late 1940s and during the 1950s, the free-piston principle created a lot of excitement among power plant designers both in Europe and the U.S. The concept promised an engine that could run on many kinds of inexpensive low grade fuels and, because it used a lot of air, had minimum pollution problems. It had a good thermal efficiency—in the 32-36% range—and was very reliable.

Typically, a free-piston engine uses freely moving pistons—not attached to crankshafts or other mechanical devices—installed in a relatively large cylinder. The most generally studied free-piston engine used two opposed and synchronized pistons in each power cylinder. Each of these is actually a dual piston which consists of a thin, wide-diameter section attached to a small-diameter, long piston section. Each piston system, then, is T-shaped. The long sections move inside a chamber in the center of the cylinder with their outer ends extending into a larger diameter cylinder for the wide end of the T.

The free-piston action begins when a fuel injector sprays fuel into the smaller piston chamber to mix with air from intake ports on one side of the chamber. The mixture ignites, driving the T-shaped pistons outward in opposite directions. As this occurs, the larger pistons in the outer cylinder chambers push against air trapped in a "bounce space" at the ends of these compression cylinders. The force developed on the ends of the smaller power pistons can result in a compression of the outer air pockets to as much as 50 times atmospheric pressure. The compressed-air cushion eventually stops the outward movement of the pistons and pushes them back with great force to start a new cycle. During these movements, series of intake and exhaust ports are covered or uncovered while various valves are opened or closed to regulate inflow of air or outflow of exhaust products. The compression chamber has a connection to the power chamber so that some of the high pressure air can be ducted into the latter. This air increases the fuel-air mixture density in the combustion chamber and thus increases energy generated by the explosion. The connection of compressed air between the two chambers also serves to control engine speed. By reducing the amount of air in the "bounce space," for instance, the amount of compression that occurs is lowered so that "bounce back," and hence engine speed, is reduced.

Returning to the first part of the cycle, when the smaller pistons move outward from the ignited fuel mixture, one thing that takes place is the uncovering of a series of exhaust ports by the end of one power piston. The high pressure, heated gases rush out these ports and into a chamber containing a series of turbine blades. As in a gas turbine engine, the exhaust gases spin the turbines. These in turn rotate the output shaft to transmit power to the wheels.

By the early 1950s, small free-piston air compressors were in production in France, and several French minesweepers were built using the concept. In 1956, General Motors installed what is believed to be the first automobile free-piston engine in a sports car design designated the XP-500. The engine for the car, called the Hyprex 4-4, used two power cylinders to develop a maximum output of 250 horsepower. The car ran well enough, but it seemed to require more parts than a conventional gasoline engine and also was much noisier. The engine's cost, GM decided, didn't promise any advantage in price over regular internal combustion designs, and little further work was done on an automobile version.

The system did seem promising for other purposes. In fact, a system of six free-piston gasifiers was developed for the merchant ship *William Patterson*. The 6,000-horsepower system worked well during the several years the *Patterson* was operated worldwide. However, though free pis-

This test car, the XP-500, was powered by a 250-horse-power free piston engine.

tons might well provide improved power for ships, the industry showed little interest in working on the new technique.

There are still some free-piston engines made in Europe. Though it has not been revived thus far for cars, the pollution situation or new technical advances in free-piston materials or parts might someday bring it back into the running.

A number of research groups have carefully studied systems that provide the direct conversion of heat to electricity. Some systems based on this method have been widely used in space vehicles or nuclear equipment. At present they have drawbacks that make their immediate use in ground vehicles unlikely. A prime example is the solar cell, which has converted sunlight directly to electricity. This has been used to power all kinds of spacecraft including those that have traveled from 30 to 150 million miles through space and have returned man's first close-in data of his planetary neighbors, Mars and Venus.

For the self-contained needs of a spacecraft, solar cells have proven to be one of the most efficient ways of generating electric power. But solar cells are very inefficient compared to the other power sources mentioned in earlier pages. Today's best cells convert into electricity only about 11 or 12% of the sunlight striking them, and each cell can only provide a minute amount of power. It takes arrays of cells numbering in the tens of thousands to provide power for a space ship the size of a car. The cost of such an array, as for the silver-zinc battery, is astronomical compared to current car costs. In addition, of course, solar cells aren't able to work when there's no sun, so special energy storage systems would be needed to build up a supply of energy for use at night or on cloudy days.

However, one or two solar-cell-powered cars have been

built, showing that the principle can work. (One of these, made by International Rectifier Corp., Inglewood, Cal., had a roof completely covered with the cells.) Someday, a breakthrough in solar cell materials with vastly improved energy conversion properties might make this system practical for a car of tomorrow.

Two other direct heat-to-electricity methods that are gaining some attention for cars of tomorrow are those based on thermoelectric power and on the science of thermionics. Thermoelectricity is a phenomenon known to scientists since 1821. In that year, T. J. Seebeck found that when two metals were joined together and a different temperature was applied to each, an electric current would flow in the loop from one to the other. The current was the result of different voltages generated at the junctions of the metals, with the voltage difference proportional to the variation in temperatures. In 1834, another scientist, J.C.A. Peltier, made a further discovery. He found that during the thermoelectric process, one of the junctions would tend to become hotter and the other colder.

The Peltier effect suggested the concept might be used in a refrigeration system. However, for over a century no progress was made in this direction because the electrical efficiency using conventional materials was very small— under one per cent. The principle was applied for certain kinds of temperature-measuring devices. After World War II, the picture changed, thanks to the breakthroughs in electronic materials that started with the discovery of the replacement for the vacuum tube, the transistor. Transistors are made from special kinds of materials called semiconductors. As semi-conductor technology grew, scientists took a new look at thermoelectricity, using these new materials to form the junction instead of conventional metals. The new systems provided efficiencies of from 4 to 5%.

This efficiency isn't good enough to provide the high electric current flow needed for practical car power systems. But it is a good enough improvement to make some devices based on Seebeck-Peltier principles practical in electronic uses.

Small thermoelectric generators are put in special refrigerators used to keep equipment cool in laboratories or as part of electronic devices that work best at low temperatures. For cases where small amounts of power are needed continuously from a reliable, simple power source, thermoelectric cells have been applied. These include systems that work automatically in remote regions, such as Arctic weather stations, special navigation buoys placed in sea lanes, and in some satellite systems.

Most of the automobile company research laboratories have been examining thermoelectric systems for possible use in future cars. Their main goal has been to develop materials with efficiencies above 5%, a target that has not been met to date. Workable air conditioners have been made in the laboratory as possible replacements for current car units. The thermoelectric systems have the advantage of reliability, since systems can be built that can cool interiors or supply accessory electric power without the need for moving parts. However, tests to date have not resulted in designs that would do this at a cost low enough to make them preferable to conventional devices. A breakthrough in the future might change this, making possible compact thermoelectric power supplies that might be built into the side panels of cars, fit between small sections of the car roof area, etc.

A system which has considerably higher power levels, based on current technology is thermionic power. The principles involved were noted in 1883 by the great Thomas Edison, inventor of the electric light bulb and many other wonders of modern technology. During tests with some of his early incandescent lamps, Edison observed that an electric current developed between the hot lamp filament and a much cooler metal structure a short distance from the filament. What was happening in this "Edison effect," he realized, was that some of the heat from the filament was being directly converted into electricity.

For a long time afterward, not too much was done to use this thermionic power for a power-generating system. Again, the development of the new electronics during and after World War II focused attention on many discoveries that previously had been almost ignored. Thermionics came in for renewed study particularly because of its potential use for generating electricity from nuclear heat sources. This was a possibility being considered for future deep space probes. By designing thermionic devices that could be installed along the periphery of a nuclear heat supply, scientists realized they might have a simple way of running the many electronic systems used by such a spacecraft.

One example of a nuclear thermionic converter was that devised by General Motors Research Laboratories for the Office of Naval Research. The converter's operation was based on the fact that a plasma* could be converted by using neutrons from the reactor to turn the atoms of a noble (chemically inactive) gas, such as neon or argon, into charged particles called ions. The plasma was introduced into the space between the hot filament of the thermionic device, called an emitter, and the cold plate, or collector. The plasma, as in the case of the electrolyte of battery systems, provided a path of high electrical current flow from emitter to conductor.

Heat from the nuclear particles simultaneously produced a plasma not only in the noble gas between the hot

---

* A gas whose atoms have been caused to ionize, or lose electrons.

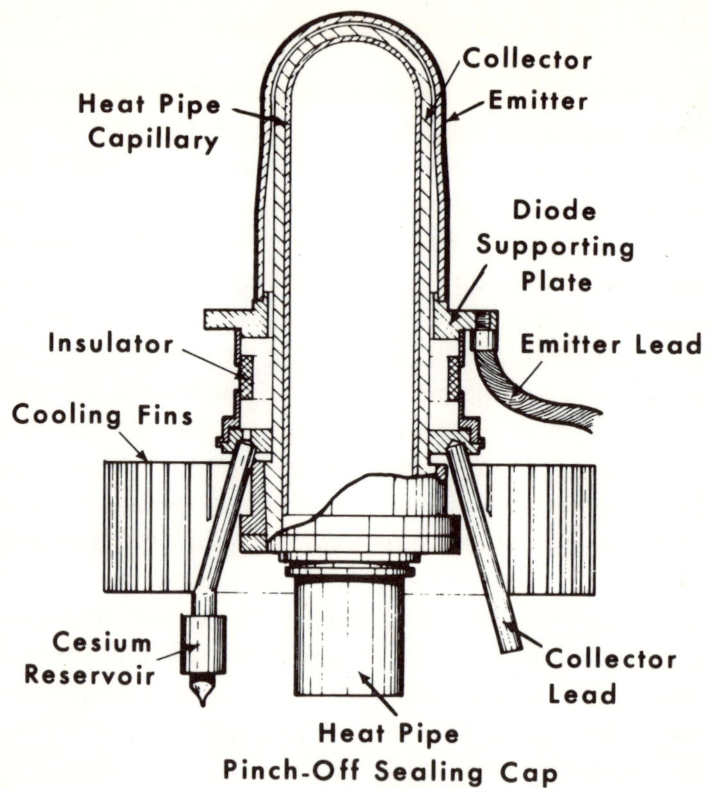

Cross section through a thermionic diode developed for General Motors Laboratories in the 1960s by Thermo Electron Corporation, Waltham, Massachusetts. The diode is heated by a fossil fuel. When its emitter is at 2552 F. and the collector at 932 F., it generates 130 amperes at 0.4 volts.

and cold junctions, but on the surface of the emitter as well. The tests showed the system could provide a very high electrical output and that the plasma in this particular case was non-corrosive. That is, the plasma didn't chemically react with other parts of the diode to eat them away. Unfortunately, the studies also showed that the high heat generated by the radiation products used up the emitter so rapidly that the diodes would only work for a relatively short time without losing power.

Since then, GM has not done much more work on spacecraft thermionics. Many other companies and research groups supported by the Atomic Energy Commission have continued to examine many variations of the thermionic power system, using hundreds of different arrangements. Thus a long-life, practical, high-voltage system may be developed in the future.

However, as Edison's early studies showed, the high heat input of a nuclear source isn't demanded for a thermionic system. Any kind of heat supply that can generate a hot enough filament temperature can provide a heat-to-electricity conversion system. Thus Westinghouse, General Electric and most battery companies as well as car builders have built low power, experimental devices using various flame sources to start the thermionic action. Eventually, some scientists and engineers hope they can develop practical thermionic systems that can provide the main power to run an electric car rather than just to serve as an emergency source.

Several methods aimed at longer range possibilities for generating power bear the impressive designations of "magnetohydrodynamics" and "electrogasdynamics." Both involve the direct generation of electricity based on the use of a flowing gas plasma. In the magnetohydrodynamic system, the plasma generates electricity by flowing through a region surrounded by a magnetic coil. The interaction between the plasma and the magnetic field causes an electric current to flow in a series of wire coils. Electrogasdynamics uses a method in which the charged particles of the gas are separated into positive ions at one electrode (called a collector electrode) and negative electrons at a second electrode (called a corona electrode). This second system might be compared to the corona effect that results around a lightning rod during an electrical storm.

At present, magnetohydrodynamics requires plasma temperatures too high for any feasible car system. However, it has been proven out as a source of high voltage electricity for large electric utilities, and test plants have been built and operated in a number of countries. Avco Corporation, Everett, Mass., pioneered a considerable amount of work in magnetohydrodynamics, including operation of the first small scale electric generator in 1959. Soviet Russia has since devoted more research and development effort to this work, with the result that the first full size magnetohydrodynamic electric generating plant was completed outside Moscow by 1972.

The fact that large voltage MHD plants are in operation is at least a guideline for future lower voltage work in this area. A breakthrough in generation and containment of high energy plasmas on a smaller scale could bring MHD into play as a future competitor for automotive power.

The electrogasdynamic method generates much smaller voltages than MHD, but they are still higher than other direct electrical conversion methods. For example, a self-sustaining converter built in General Motors Laboratories provides an output of 40,000 volts from an electrically charged gas. The system built at GM looks something like a king-sized doctor's stethoscope or a giant wishbone, with two roughly six-inch diameter spheres at the ends of the two tubular arms. In operation, gas compressed to a high pressure comes in at the head of the wishbone where it flows past a wire that is at a very high voltage. This voltage changes the atoms of the neutral gas to ions, some of which are converted to electrons by the attractor electrode located around the inside walls of the tube section surrounding the high voltage wire. The ionized particles are attracted to the wire.

Though the positively charged particles try to move towards the corona electrode, many of them are prevented

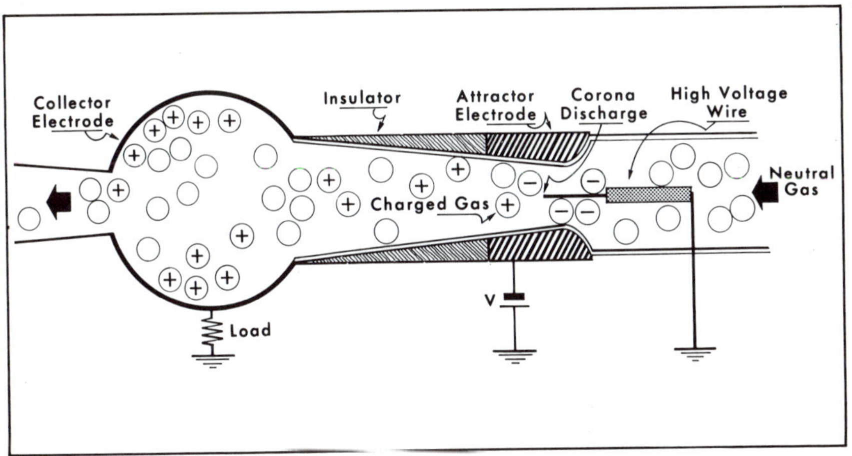

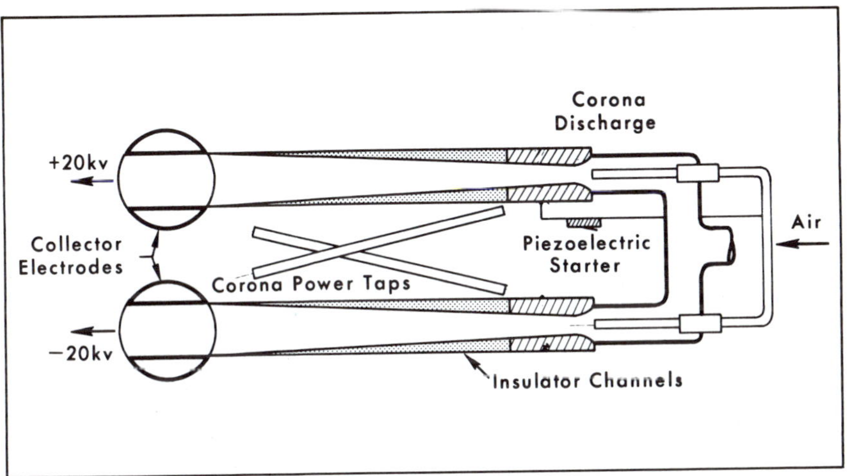

The top diagram shows what happens in an Electrogas dynamics converter. High velocity gas flows through a high voltage discharge. The discharge causes many of the neutral atoms to ionize, or lose electrons. Later on, separation of positive ions at the collector electrode and negative electrons at the corona electrode causes direct conversion of gas kinetic energy to electricity. The lower diagram shows the application of these principles in a General Motors EGD system that can generate 40 kilovolts from an electrically charged gas.

from reaching it by the high speed of the incoming gas. These positive particles are swept down towards the spheres where they collect on a smooth collector section. As these particles build up on the inner sphere surfaces, they create a back pressure tending to interfere with the flow of neutral atoms. The continued interaction results in a buildup of electrical energy on the collector that appears as an electrical voltage between the collector electrode and the "ground." (The word ground is an electrical term that refers to a wire connected to a material that will safely get rid of excess electricity. This is a requirement in any electrical circuit to prevent a buildup of energy that might burn out the main system or cause a short circuit.) In the electrogasdynamic system, the electrical voltage generated in this way would first be fed through the device that is to be operated, such as an electric motor, then from there sent to the ground.

The GM system uses a special electromechanical device called a transducer to start the system running. A transducer is a device that can change mechanical energy into electrical energy (or vice versa). This is done by vibrating the transducer's atoms so that either a current is made to flow in an electric wire; or, conversely, the vibration sends out pressure pulses into a medium such as air or a gas. An example of the use of a transducer is the conventional telephone. Here, the pressure waves in the air caused by the voice make a transducer in the mouthpiece vibrate, an action that sets up an equivalent electrical wave in the phone wire.

In the EGD unit, the transducer is vibrated to trigger a corona discharge in the upper half of the converter. This action generates a high voltage pulse on the upper attractor electrode. Part of this voltage is then directed through wires to set off the corona discharge in the lower generator, which then helps reactivate the upper electrode.

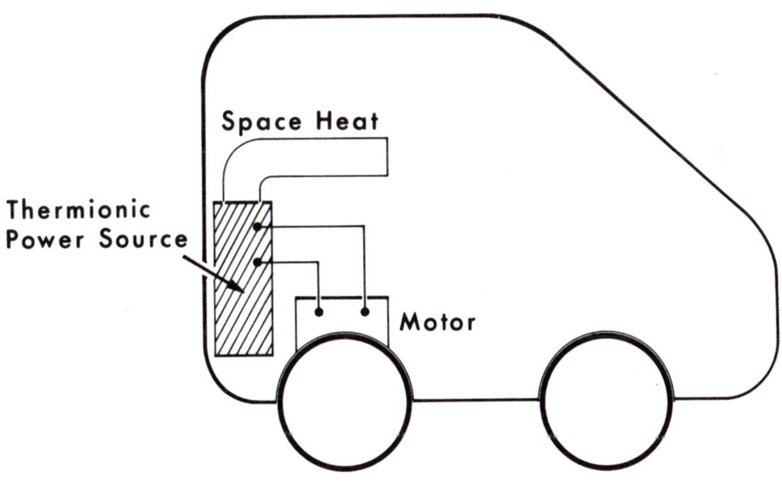

Simplified diagram shows the possible installation of a thermionic system in a car of the future. A practical thermionic engine for a small electric car might contain 20 diodes, each generating 50 watts of electricity at an assumed over-all efficiency of 5%.

As a result, once the transducer starts this "Ping-pong" operation, the converter keeps working until it is turned off.

The EGD technique is receiving considerable research attention because it promises a number of eventual uses, from the primary power to run an electric car, to such other things as running an ignition system or providing pollution control by precipitating out chemical particles from the car exhaust.

Unfortunately, the system still has many problems. The biggest drawback is very low efficiency at present. The system can develop very high voltages, but most of this energy is wasted—only a small part can be converted to electric current to perform useful work. For instance, the EGD arrangement just described takes in air at 50-pounds per-square-inch pressure which has a calculated kinetic energy

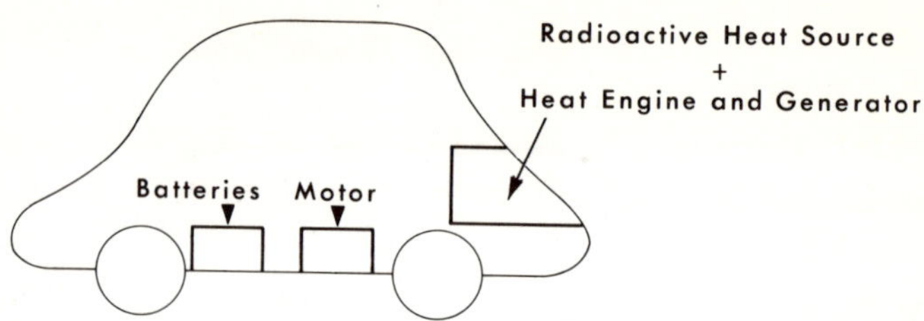

Schematic of a hybrid vehicle with a radioactive heat source.

Radioisotope	Waste Fission Products	Plutonium	Curium	Thallium	Promethium
Half-life (years)	30	89	18	4	3
Total Weight of Heat Device (lb)	2,400	50	500	300	300
Initial Cost/Car (millions of $)	0.1	2.7	0.3	0.5	1.0
U.S. Supply in 1980 (kw per year)	2,500	100	1	100	100

Selected Radioisotope Heat Sources

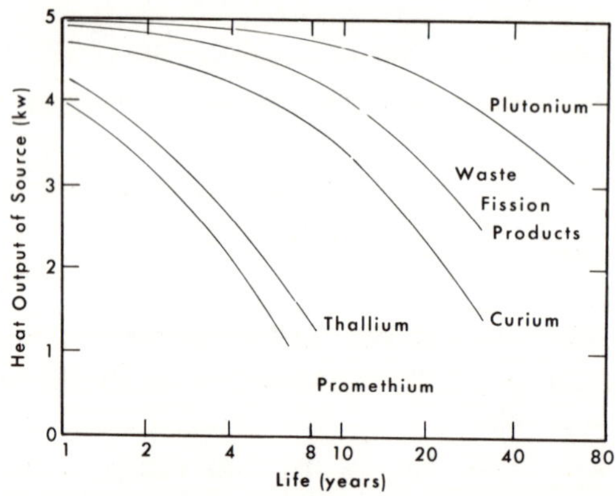

Life Characteristics of Radioisotope Heat Sources

A hybrid car may someday be built using a radioactive heat source as shown in the schematic at top. The table and chart compare some of the properties of currently available radioisotope heat sources.

of 6,600 watts. The power output at 40 kilovolts is only 1.4 watts, or an overall efficiency of 0.02%. In the future, though, improvements in materials or the arrangment of system parts may upgrade the efficiency to a practical level.

The last system to be reviewed here is one in which nuclear energy might be used to power a future car. Obviously, a very small amount of radioactive material could provide energy for a car for a very long time. But even a small amount of relatively low radioactivity requires a large amount of special shielding material to prevent dangerous particles from reaching the car passengers or escaping into the air. If the required shielding *could* be developed, a nuclear car using energy from the nuclear fuel to operate an electric motor, battery, etc., could be completely pollution free.

Studies have shown that radioisotopes such as plutonium-210, curium, thallium or ruthenium could be used in an automotive system. They can be safely contained, can turn out the needed power density, and have a radioactive lifetime of from a few years to as much as 80 years. However, their cost, based on present technology, would be astronomical—anywhere from $100,000 to almost $3 million a car, and they would still be very heavy, with the necessary shielding weighing roughly 2,400 lbs. per car.

Despite their shortcomings, none of these advanced power concepts can be ruled out. Continued research may bring one or more of these systems or modifications of them to the fore as the vitally needed new car engine. Then again, the eventual successor to the internal combustion engine may be something unlike any power system covered in this book. It may exist today as only the glimmer of an idea in the fertile brain of a practicing engineer or scientist, or in the mind of a youngster whose future as an inventor of automobile engines may lie decades ahead of him or her.

Index

ABOUT THE AUTHOR

IRWIN STAMBLER received his degree in aeronautical engineering from New York University and worked for a number of years in the aviation industry. Combining his engineering knowledge and his writing ability, he began free-lance writing for magazines and eventually became the engineering editor of *Space/Aeronautics*, a leading trade journal for the aviation and space industry. At present, he is Western Editor for *Industrial Research Magazine* and co-publisher of Technology Forecasts newsletter. He has written a number of books on scientific subjects for young people and, in addition, a book for adults about undersea exploration. Mr. Stambler lives with his wife and four children in Beverly Hills, California.